Manual for
QUALITY CONTROL
For Plants and Production of
ARCHITECTURAL PRECAST CONCRETE PRODUCTS

THIRD EDITION
MNL-117-96

PRECAST/PRESTRESSED CONCRETE INSTITUTE

175 W. JACKSON BLVD.
CHICAGO, IL 60604
PHONE: 312/786-0300
FAX: 312/786-0353
e-mail: info@pci.org

Copyright © 1996 by
PRECAST/PRESTRESSED CONCRETE INSTITUTE

All rights reserved. No part of this book may be reporduced in any form without permission in writing from the publisher. Printed in the U.S.A.

First Edition, 1968
Second Edition, 1977
Second Edition, Third Printing, 1984
Second Edition, Fourth Printing, 1988
Third Edition, 1996

ISBN 0-937040-48-7

FOREWORD

This Manual has been prepared as a guideline for quality assurance of architectural precast concrete. Since the products are custom designed, the many combinations of shape, size, color and texture require a great degree of craftsmanship. Therefore it is important to implement and maintain the quality control Standards as given in this Manual to achieve the specific performance and aesthetic requirements of a project.

Materials and performance requirements for the architectural precast concrete should be clearly stated in the plans and specifications. These requirements should neither be open to interpretation nor unduly restrictive for the project, but should be written to conform with the intended use of the architectural precast concrete. Personnel in the manufacturer's organization should be thoroughly trained and competent in order to achieve quality architectural precast concrete products.

The first edition of the Manual for Quality Control for Plants and Production of Architectural Precast Concrete Products was prepared by the PCI Plant Certification Subcommittee for Plant Production of Architectural Precast Concrete Products. Subsequent to the publication and use of the first edition, a plant certification program was established for the precast and prestressed concrete industry. The inspection of architectural precast concrete production facilities, carried out under this program, was based on the recommended practices and criteria outlined in that Manual. Experience by both the manufacturers and the inspection teams led to the second edition in 1977.

The third edition which is even more demanding of a high standard of industry practice was prepared by the PCI Architectural Precast Concrete Services Committee and the PCI Plant Certification Committee. It represents state-of-the-art procedures and is the industry standard for achieving consistently high quality. Committee members working on this Manual were as follows:

ARCHITECTURAL PRECAST CONCRETE SERVICES COMMITTEE

Donald G. Clark, Chairperson

Kevin Anderson	Tom Kelley
James T. Engle	Edward S. Knowles
Donald L. Faust, Sr.	Douglas L. Lorah
Charles L. Fister	Richard C. Nash
Sidney Freedman*	Richard C. Page, Jr.**
Marvin F. Hartsfield	Ned Piccinini
*Principal author	Arthur G. Salzman
**Past Chairman	

PLANT CERTIFICATION COMMITTEE

Dino J. Scalia, Chairperson

Ray Andrews, Jr.	Ted J. Gutt
John H. Bachman, Jr.	Ray L. Kennedy
Henry Clark	Joel Kessell
Theodore W. Coons	Michael W. LaNier
John S. Dick	Robert McCrossen
Larry G. Fischer	Gary E. Oakes
	Stanley J. Ruden

INTRODUCTION
MNL 117, Third Edition

> The Standard and Commentary are presented in a side-by-side column format, with the Standard text placed in the left column and the corresponding Commentary text aligned in the right column. The Standard has been printed in Helvetica, the same type face in which this paragraph is set.
>
> This paragraph is set in Times, all portions of the Commentary are printed in this type face. Commentary article numbers are preceded by a "C" to further distinguish them from Standard article numbers. The information contained in the Commentary is not part of the Standard. It shall not be used in judging quality control or production procedures. It does provide suggestions to help in carrying out the requirements or intent of the standard.

Architectural precast concrete panels, through the application of finish, shape, color or texture, contribute to the architectural form and finished effect of a structure. Not generally included are the so-called industrialized precast products (standard shapes), such as double or single tees, channel sections, and flat or hollow-core slabs usually produced in fixed, long-line forms. Architectural precast concrete units may be manufactured with conventional mild steel reinforcement, or they can be prestressed. Design flexibility in surface appearance is possible by incorporating various textures and finishes and through the use of different cements, coarse and fine aggregates, and pigments into the concrete mix. Natural stone or clay products may be used as a veneer finish or alternatively, panels may be painted or stained to achieve the required colors.

The architect/engineer is directed to Appendix B for a listing of the responsibilities which are to be considered in the preparation of plans and specifications for an architectural concrete project.

This manual is divided into two parts. The first part contains Divisions 1 through 7 which form the basis for PCI Plant Certification in product Group A1 — architectural precast concrete products. It is conformance to these Standards which is audited during each PCI plant inspection and provides the criteria for evaluation of the plant's capabilities.

The final part — Appendices — contains summaries of useful information for both the manufacturer and specifier.

The Standard is intended to be used as the basis for a quality assurance program for the manufacture of architectural precast concrete products and to provide requirements for quality production practices. The Standard portion is intended to serve as a specification reference document. It should be augmented as required for specific operations and products by the specifier or producer. The Commentary provides amplification and explanation of the Standard to assist the manufacturer.

Routine conformance to requirements of this Standard should result in products of consistent and optimum quality if used in concert with proven and documented plant-specific procedures. Optimum quality is defined as that level of quality, in terms of appearance, strength and durability, which is appropriate for the specific product and its particular application. In other words, it is that quality which is both economical to achieve and suitable for the particular purpose it serves as a component in the overall project.

This Standard provides a minimum level of quality, but there is no intent to place a ceiling on excellence. The degree of success in specifying and obtaining optimum quality for products will depend on the combined efforts of designers and manufacturers to define and coordinate their individual requirements, responsibilities and expectations.

No Manual of this type can be all-inclusive. The requirements and recommendations given herein are a general presentation of the important factors governing the quality of architectural precast concrete. Their value is dependent on rational application and a determination on the part of the individual producer to establish a standard of quality that will be recognized and respected by the specifier.

Quality assurance begins when the architect determines shape, size, color and texture for the architectural precast concrete products for a specific project. These characteristics may then determine the methods of manufacture, as well as the handling and installation techniques. Consultation with qualified representatives of experienced manufacturers will be of great value in achieving high quality products at a reasonable cost to the owner.

The Standard indicates the requirements to obtain the desired quality but not the means or methods. It is not the intention of the Manual to restrict individual plant techniques. For example, a manufacturer's methods for mixing, placement, consolidation and curing of concrete will be acceptable, provided these methods can consistently result in uniform and durable concrete of the specified quality.

The Commentary contains suggestions to help in carrying out the requirements or intent of the Standard.

This Manual has been prepared on the basis of current good practice. As significant changes in materials or process technology occur, revisions will be made to this Manual.

Note: The production of architectural precast concrete may involve hazardous materials, operations, and equipment. This manual does not purport to address the safety problems associated with production. It is the responsibility of the plant to establish appropriate safety and health practices. The plant should determine the applicability of any regulatory limitations.

TABLE OF CONTENTS

Foreword iii
Introduction v
Definitions xi

DIVISION 1 — QUALITY SYSTEM

1.1 Objective

1.2 Plant Quality Assurance Program
1.2.1 General 1
1.2.2 Documented Procedures 2
1.2.3 Management Responsibilities 3

1.3 Personnel
1.3.1 General 4
1.3.2 Engineering 4
1.3.3 Drafting 4
1.3.4 Production 5
1.3.5 Quality Control 5

1.4 Design Responsibilities
1.4.1 General 5
1.4.2 Shop Drawings 6

1.5 Project Samples
1.5.1 General 6
1.5.2 Size and Shape 6
1.5.3 Identification 7
1.5.4 Visual Mockups and Initial Production Approval of Architectural Finishes 7

DIVISION 2 — PRODUCTION PRACTICES

2.1 General Objectives and Safety
2.1.1 General 9
2.1.2 Plant Safety 9

2.2 Production and Curing Facilities
2.2.1 Area Requirements 9
2.2.2 Mold Fabrication 10
2.2.3 Storage of Release Agents and Retarders 10
2.2.4 Hardware Fabrication and Storage 10
2.2.5 Casting Area and Equipment 11
2.2.6 Curing and Finishing Areas 12
2.2.7 Handling Equipment 12
2.2.8 Storage Area for Finished Products 13

2.3 Welding
2.3.1 Structural Steel 13
2.3.2 Reinforcement 16
2.3.3 Stud Welding 20

2.4 Molds
2.4.1 Materials and Construction 21
2.4.2 Verification and Maintenance 24

2.5 Hardware Installation 24

2.6 Product Handling
2.6.1 General 27
2.6.2 Stripping 28
2.6.3 Yard Storage 28
2.6.4 Cleaning 29
2.6.5 Loading 29

2.7 Surface Finishes
2.7.1 General 30
2.7.2 Smooth 31
2.7.3 Sand or Abrasive Blast 32
2.7.4 Acid Etched 33
2.7.5 Retarded 34
2.7.6 Tooled or Bushhammered 35
2.7.7 Honed or Polished 35
2.7.8 Form Liner 36
2.7.9 Veneer Facing Material 36
2.7.10 Sand Embedded Materials 39
2.7.11 Unformed Surface Finishes 40
2.7.12 Applied Coatings 41

2.8 Repairs 41

2.9 Acceptability of Appearance 42

2.10 Sealers or Clear Surface Coatings 44

DIVISION 3 — RAW MATERIALS AND ACCESSORIES

3.1 Concrete Materials
3.1.1 General 47
3.1.2 Cement 47
3.1.3 Facing Aggregates 47
3.1.4 Backup Aggregates 49
3.1.5 Aggregates for Lightweight Concrete 50
3.1.6 Mixing Water 50
3.1.7 Admixtures 51

3.2 Reinforcement and Hardware

3.2.1	Reinforcing Steel 53		4.11	Scale Requirements 79
3.2.2	Prestressing Materials 55		4.12	Requirements for Water Measuring Equipment 80
3.2.3	Hardware and Miscellaneous Materials 58		4.13	Requirements for Batchers and Mixing Plants
3.2.4	Handling and Lifting Devices 62		4.13.1	General 81
3.3	Insulation 63		4.13.2	Requirements for Concrete Mixers 82
			4.13.3	Mixer Requirements 82
3.4	Welding Electrodes 63		4.13.4	Maintenance Requirements for Concrete Mixers 83

DIVISION 4 — CONCRETE

4.1	Mix Proportioning		4.14	Concrete Transportation Equipment
4.1.1	Qualification of New Concrete Mixes 67		4.14.1	General 83
4.1.2	Specified Concrete Strength 68		4.14.2	Requirements for Concrete Agitating Delivery Equipment 83
4.1.3	Statistical Concrete Strength Considerations 68			
4.1.4	Proportioning to Ensure Durability of Concrete 69		4.15	Placing and Handling Equipment 83
4.2	Special Considerations for Air Entrainment 70		4.16	Batching and Mixing Operations
			4.16.1	General 84
			4.16.2	Batching of Aggregates 85
4.3	Compatibility of Face and Backup Mixes 71		4.16.3	Batching of Cement and Pigments 86
			4.16.4	Batching of Water 86
			4.16.5	Batching of Admixtures 86
4.4	Proportioning for Appearance of Concrete Surfaces 71		4.17	Mixing of Concrete
			4.17.1	General 88
4.5	Mix Proportioning for Concrete made with Structural Lightweight Aggregate 71		4.17.2	Methods of Concrete Mixing 88
			4.17.3	Mixing Time and Concrete Uniformity ... 89
			4.17.4	Mixing Time - Stationary Mixers 90
			4.17.5	Mixing Time - Shrink Mixing 90
4.6	Proportioning for Concrete Workability 72		4.17.6	Mixing Time - Truck Mixing 90
			4.17.7	Special Batching and Mixing Requirements for Lightweight Aggregates 90
4.7	Water-Cement Ratio			
4.7.1	General 72		4.17.8	Cold Weather Mixing 91
4.7.2	Relationship of Water-Cement Ratio to Strength, Durability and Shrinkage 73		4.17.9	Hot Weather Mixing 92
4.7.3	Relationship of Water-Cement Ratio to Workability 73		4.18	Requirements for Transporting and Placing of Concrete
			4.18.1	General 92
4.8	Effects of Admixtures 73		4.18.2	Transporting and Placing Concrete 93
			4.18.3	Preventing Aggregate Segregation 93
4.9	Storage and Handling of Concrete Materials		4.18.4	Preparation of the Molds 94
			4.18.5	Placing Concrete Under Severe Weather Conditions 94
4.9.1	General 74			
4.9.2	Storage and Handling of Aggregates 74		4.18.6	Placing Concrete in Wet and Rainy Conditions 94
4.9.3	Storage and Handling of Cement 75			
4.9.4	Storage and Handling of Admixtures and Pigments 76		4.18.7	Placing Concrete in Hot or Windy Conditions 95
			4.18.8	Placing Concrete in Cold Weather Conditions 96
4.10	Batching Equipment Tolerances 78		4.18.9	Placing Facing Concrete 96
			4.18.10	Placing Backup Concrete 97

4.19 Consolidation of Concrete
- 4.19.1 General 97
- 4.19.2 Consolidation of Lightweight Concrete 98
- 4.19.3 Consolidation of Face and Backup Mixes 98
- 4.19.4 Use of Internal Vibrators 98
- 4.19.5 Use of External Vibrators 99
- 4.19.6 Use of Surface Vibrators 99
- 4.19.7 Use of Vibrating Tables 100

4.20 Requirements for Curing Concrete
- 4.20.1 General 100
- 4.20.2 Curing Temperature Requirements 100
- 4.20.3 Curing to Attain Specified Stripping or Release Strength 101

4.21 Accelerated Curing of Concrete
- 4.21.1 General 101
- 4.21.2 Curing with Live Steam 103
- 4.21.3 Curing with Radiant Heat and Moisture 103

4.22 Curing by Moisture Retention without Supplemental Heat
- 4.22.1 General 103
- 4.22.2 Moisture Retention Enclosure 104
- 4.22.3 Curing with Membrane Curing Compound 104

DIVISION 5 — REINFORCEMENT AND PRESTRESSING

5.1 Reinforcing Steel
- 5.1.1 General 105
- 5.1.2 Storage of Reinforcing Steel 105
- 5.1.3 Fabrication of Reinforcing Steel 106
- 5.1.4 Installation of Reinforcing Steel 107

5.2 Tensioning
- 5.2.1 General Tensioning Requirements 110
- 5.2.2 Tensioning of Tendons 110
- 5.2.3 Methods of Force Measurement 111
- 5.2.4 Gauging Systems 112
- 5.2.5 Control of Jacking Force 112
- 5.2.6 Wire Failure in Strand or Tendon 112
- 5.2.7 Calibration Records for Jacking Equipment 113

5.3 Pretensioning
- 5.3.1 Storage of Prestressing Steel 113
- 5.3.2 General 114
- 5.3.3 Strand Surfaces 114
- 5.3.4 Stringing of Strands 115
- 5.3.5 Strand Chucks and Splice Chucks 115
- 5.3.6 Strand Splices 116
- 5.3.7 Strand Position 116
- 5.3.8 Initial Tensioning 116
- 5.3.9 Measurement of Elongation 117
- 5.3.10 Elongation Calculation and Corrections 117
- 5.3.11 Force Corrections 119
- 5.3.12 Final Stressing of Strands 120
- 5.3.13 Detensioning 120
- 5.3.14 Protection of Strand Ends and Anchorages 121

5.4 Post-Tensioning of Plant-Produced Products
- 5.4.1 General 122
- 5.4.2 Details and Positions for Ducts 123
- 5.4.3 Friction in Ducts 124
- 5.4.4 Tensioning 124
- 5.4.5 Anchorages 124
- 5.4.6 Grouting 125
- 5.4.7 Sealing of Anchorages 125

DIVISION 6 — QUALITY CONTROL

6.1 Inspection
- 6.1.1 Necessity for Inspection 127
- 6.1.2 Scope of Inspection 127

6.2 Testing
- 6.2.1 General 128
- 6.2.2 Acceptance Testing of Materials 129
- 6.2.3 Production Testing 138
- 6.2.4 Special Testing 153

6.3 Records
- 6.3.1 Recordkeeping 153
- 6.3.2 Suppliers' Test Reports 154
- 6.3.3 Tensioning Records 155
- 6.3.4 Concrete Records 156
- 6.3.5 Calibration Records for Equipment 156

6.4 Laboratory Facilities
- 6.4.1 General 157
- 6.4.2 Quality Control Testing Equipment 157
- 6.4.3 Test Equipment Operating Instructions 157

DIVISION 7 — PRODUCT TOLERANCES

7.1 Requirements for Finished Product
- 7.1.1 Product Tolerances – General 159
- 7.1.2 Product Tolerances 160

APPENDICES

APPENDIX A
Guidelines for Developing Plant Quality System Manual 167

APPENDIX B
Design Responsibilities and Considerations 175

APPENDIX C
Finish Samples 179

APPENDIX D
Chuck Use and Maintenance Procedure 181

APPENDIX E
Sample Record Forms 183

APPENDIX F
PCI Plant Certification Program 189

APPENDIX G
Reference Literature 191

APPENDIX H
Sample Tensioning Calculations 197

APPENDIX I
Erection Tolerances 207

APPENDIX J
Architectural Trim Units 209

INDEX 215

DEFINITIONS

Accelerated curing — See Curing.

Admixture — A material other than water, aggregates and cement used as an ingredient in concrete, mortar or grout to impart special characteristics.

Aggregate — Granular material, such as sand, gravel, and crushed stone used with a cementing medium to form a hydraulic-cement concrete or mortar.

Aggregate, structural lightweight — Aggregate with a dry, loose weight of 70 lbs/ft^3 (1121 kg/m^3) or less.

Air entraining admixture — A chemical added to the concrete for the purpose of providing minute bubbles of air (generally smaller than 1 mm) in the concrete during mixing to improve the durability of concrete exposed to cyclical freezing and thawing in the presence of moisture.

Ambient temperature — The temperature of the air surrounding the forms and molds into which concrete is to be cast, or of the air surrounding an element during curing.

Anchorage — The means by which the prestressing force is permanently transmitted from the prestressing steel to the concrete. In post-tensioned applications, a mechanical device comprising all components required to anchor the prestressing steel and transmit the prestressing force to the concrete.

Architectural precast concrete — A product with a specified standard of uniform appearance, surface details, color, and texture.

Architectural precast concrete Trim Units — Wet cast products with a high standard of finish quality and of relatively small size that can be installed with equipment of limited capacity, such as sills, lintels, coping, cornices, quoins, medallions, bollards, benches, planters, and pavers.

Backup mix — The concrete mix cast into the mold after the face mix has been placed and consolidated.

Bleeding — A form of segregation in which some of the water in a mix rises to the surface of freshly placed concrete; also known as water gain.

Blocking — Materials used for keeping concrete elements from touching each other or other materials during storage and transportation.

Bondbreaker — A substance placed on a material to prevent it from bonding to the concrete, or between a face material such as natural stone and the concrete backup.

Bonding agent — A substance used to increase the bond between an existing piece of concrete and a subsequent application of concrete such as a patch.

Bull float — A tool comprising a large, flat, rectangular piece of wood, aluminum, or magnesium usually 8 in. (200 mm) wide and 42 to 60 in. (1.0 to 1.5 m) long, and a handle 4 to 16 ft. (1 to 5 m) in length used to smooth unformed surfaces of freshly placed concrete.

Bugholes — Small holes on formed concrete surfaces formed by air or water bubbles, sometimes called blowholes.

Camber — (1) The deflection that occurs in pre-stressed concrete elements due to the net bending resulting from application of a pre-stressing force. (It does not include dimensional inaccuracies); and (2) A built-in curvature to improve appearance.

Certification — Assurance by a competent third party organization, operating on objective criteria and which is not subject to undue influences from the manufacturer or purchaser or to financial considerations, that elements are consistently produced in conformity with a specification. It not only proclaims compliance of a product with a specification, but also that the manufacturer's quality control arrangements have been approved and that a continuing audit is carried out.

Clearance — Interface space (distance) between two items.

Coarse aggregate — Aggregate predominately retained on the U.S. Standard No. 4 (4.75 mm) sieve; or that portion of an aggregate retained on the No. 4 (4.75 mm) sieve.

Compaction — The process whereby the volume of the concrete is reduced to the minimum practical space by the reduction of voids usually by vibration, tamping or some combination of these.

Connection — Device for the attachment of precast concrete units to each other or to the building or structure.

Covermeter — See R-meter.

Crazing — A network of visible, fine hairline cracks in random directions breaking the exposed face of a panel into areas of from 1/4 in. to 3 in. (6 to 75 mm) across.

Creep — The time dependent deformation (shortening) of prestressing steel or concrete under sustained loading.

Curing — The maintenance of humidity and temperature of freshly placed concrete during some definite period following placing, casting, or finishing to assure satisfactory hydration of the cementitious materials and proper hardening of the concrete; where the curing temperature remains in the normal environmental range [generally between 50 and 90 deg. F (10 and 32 deg. C)] use the term normal curing; where the curing temperature is increased to a higher range [generally between 90 and 150 deg. F (32 and 66 deg. C)] use the term accelerated curing.

Detensioning of strand or wire — The transfer of strand or wire tension from the bed anchorage to the concrete.

Draft — The slope of concrete surface in relation to the direction in which the precast concrete element is withdrawn from the mold; it is provided to facilitate stripping with a minimum of mold breakdown.

Dunnage — See Blocking.

Elastic shortening — The shortening of a member which occurs immediately after the application of the prestressing force.

Elongation — Increase in length of the prestressing steel (strand) under the applied prestressing force.

Exposed aggregate concrete — Concrete manufactured so that the aggregate on the face is left protruding.

Face mix — The concrete at the exposed face of a concrete unit used for specific appearance purposes.

Fine aggregate — Aggregate passing the 3/8 in. (9.5 mm) sieve and almost entirely passing the No. 4 (4.75 mm) sieve and predominately retained on the No. 200 (75 μm) sieve; or that portion of an aggregate passing the No. 4 (4.75 mm) sieve and predominately retained on the No. 200 (75 μm) sieve.

Form — See Mold.

Formed surface — A concrete surface that has been cast against formwork.

Form release agent — A substance applied to the mold for the purpose of preventing bond between the mold and the concrete cast in it.

Friction loss — In post-tensioned applications, the stress (force) loss in a prestressing tendon resulting from friction created between the strand and sheathing due to curvature in the tendon profile during stressing.

Gap-graded concrete — A mix with one or a range of normal aggregate sizes eliminated, and/or with a heavier concentration of certain aggregate sizes over and above standard gradation limits. It is used to obtain a specific exposed aggregate finish.

Grout — A mixture of cementitious materials and water, with or without sand or admixtures.

Hardware — Items used in connecting precast concrete units or attaching or accommodating adjacent materials or equipment. Hardware is normally divided into three categories:

    Contractor's hardware — Items to be placed on or in the structure in order to receive the precast concrete units, e.g., anchor bolts, angles, or plates with suitable anchors.

    Plant hardware — Items to be embedded in the concrete units themselves, either for connections and precast concrete erector's work, or for other trades, such as mechanical, plumbing, glazing, miscellaneous iron, masonry, or roofing trades.

    Erection hardware — All loose hardware necessary for the installation of the precast concrete units.

Homogeneous mix — A uniform concrete mix used throughout a precast concrete element.

Initial prestress — The stress (force) in the tendon immediately after transferring the prestressing force to the concrete.

Jacking force — The maximum temporary force exerted by the jack while introducing the prestressing force into the concrete through the prestressing strand.

Jig — A template or device to align parts of an assembly, usually for pre-assembling reinforcing steel and hardware cages, with a minimum of measurement to attain consistent accuracy from one cage to the next.

Laitance — Residue of weak and nondurable

material consisting of cement, aggregate fines, or impurities brought to the surface of plastic concrete by bleed water.

Lifting frame (or beam) — A rigging device designed to provide two or more lifting points of a precast concrete element with predictable load distribution and pre-arranged direction of pulling force during lifting.

Mark number — The individual identifying mark assigned to each precast concrete unit predetermining its position in the building.

Master mold — A mold which allows a maximum number of casts per project; units cast in such molds need not be identical, provided the changes in the units can be simply accomplished as pre-engineered mold modifications.

Matrix — The portion of the concrete mix containing only the cement and fine aggregates (sand).

Miter — An edge that has been beveled to an angle other than 90 deg.

Mold — The container or surface against which fresh concrete is cast to give it a desired shape; sometimes used interchangeably with form. (The term is used in this Manual for custom made forms for specific jobs while forms are used for standard forms or forms of standard cross section.)

Pattern or positive — A replica of all or part of the precast element sometimes used for forming the molds in concrete or plastic.

Plastic cracking — Short cracks often varying in width along their length that occur in the surface of fresh concrete soon after it is placed and while it is still plastic.

Post-tensioning — A method of prestressing concrete whereby the tendon is kept from bonding to the plastic (wet) concrete, then elongated and anchored directly against the hardened concrete, imparting stresses through end bearing.

Precast engineer — The person or firm who designs precast concrete members for specified loads and who may also direct the preparation of the shop drawings.

Pretensioning — A method of prestressing concrete whereby the tendons are elongated, anchored while the concrete in the member is cast, and released when the concrete is strong enough to receive the forces from the tendon through bond.

Production drawings — A set of instructions in the form of diagrams and text which contain all the information necessary for the manufacturer to produce the unit.

Quality — The appearance, strength and durability which is appropriate for the specific product, its particular application and its expected performance requirements. The totality of features and characteristics of a product that bear on its ability to satisfy stated or implied needs.

Quality assurance (QA) — All those planned or systematic actions necessary to ensure that the final product or service will satisfy given requirements for quality and perform intended function.

Quality control (QC) — Those actions related to the physical characteristics of the materials, processes, and services, which provide a means to measure and control the characteristics to predetermined quantitative criteria.

Quirk miter — A corner formed by two chamfered members to eliminate sharp corners and ease alignment.

R-meter — An electronic device used to locate and size reinforcement in hardened concrete.

Retarder — An admixture which delays the setting of cement paste and therefore of concrete.

Retarder, surface — A material used to produce exposed aggregate concrete by retarding or delaying the hardening of the cement paste on a concrete surface within a time period and to a depth to facilitate removal of this paste after the concrete element is otherwise cured.

Retempering — The addition of water or admixture and remixing of concrete which has started to stiffen in order to make it more workable.

Return — A projection which is angles away from the main face or plane of view.

Reveal — (1) Groove in a panel face generally used to create a desired architectural effect; and (2) The depth of exposure of the coarse aggregate in the matrix after production of an exposed aggregate finish.

Rustication — A groove in a panel face for architectural appearance; also reveal.

Sandwich wall panel — A prefabricated panel which is a layered composite formed by attaching two wythes or skins of concrete separated by an insulating core.

Scabbing — A finish defect in which parts of the form face including release agent adhere to the concrete, some probable causes are an excessively rough form face, inadequate application of release agent, or delayed stripping.

Scouring — Irregular eroded areas or channels with exposed stone or sand particles; some probable causes of this finish defect are excessively wet concrete mix, insufficient fines, water in form when placing, poor vibration practices, and low temperature when placing.

Sealer — A clear chemical compound applied to the surface of precast concrete units for the purpose of improving weathering qualities or reducing water absorption.

Segregation — The tendency for the coarse particles to separate from the finer particles in handling; in concrete, the coarse aggregate and drier material remaining behind and the mortar and wetter material flowing ahead; this also occurs in a vertical direction when wet concrete is overvibrated or dropped vertically into the forms, the mortar and wetter material rising to the top; in aggregate, the coarse particles roll to the outside edges of the stockpile.

Self stressing form — A form provided with suitable end bulkheads and sufficient cross-sectional strength to resist the total prestressing force.

Set-up — The process of preparing molds or forms for casting, including installation of materials (reinforcement and hardware) prior to the actual placing of concrete.

Sheathing — A material covering forming an enclosure around the prestressing steel to avoid temporary or permanent bond between the prestressing steel and the surrounding concrete.

Shrinkage — The volume change in precast concrete units caused by drying normally occurring during the hardening process of concrete.

Shop drawings — (1) Collective term used for erection drawings, production drawings and hardware details; and (2) Diagrams of precast concrete members and their connecting hardware, developed from information in the contract documents. They show information needed for both field assembly (erection) and manufacture (production) of the precast concrete units.

Specially Finished Structural Precast Concrete — A product fabricated using forms and techniques common to the production of structural elements as defined in MNL-116 and having specified surface finishes that require uniformity and detailing more demanding than the requirements of MNL-116. These surface finish requirements should be clearly specified, and verified with appropriate samples and mockups.

Spreader beam — A frame of steel channels or beams attached to the back of a panel, prior to stripping, for the purpose of evenly distributing loads to inserts and for lifting the panel about its center of gravity.

Strand — A group of wires laid helically over a central-core wire. A seven-wire strand would thus consist of six outer wires laid over a single wire core.

Strand anchor — A device for holding a strand under tension, sometimes called a strand chuck or vice.

Stripping — The process of removing a precast concrete element from the form in which it was cast.

Strongback/Stiffback — A steel or wooden member which is attached to a panel for the purpose of adding stiffness during handling, shipping and/or erection.

Structural lightweight concrete — Structural concrete made with lightweight aggregate with an air-dry unit weight of the concrete in the range of 90 to 115 lb/ft^3 (1440 to 1850 kg/m^3) and a 28-day compressive strength of more than 2500 psi (17.24 MPa).

Superplasticizer — A high range water reducing (HRWR) admixture producing concrete of significantly higher slump without addition of water.

Surface retarder — A material used to retard or prevent the hardening of the cement paste at a concrete surface to facilitate removal of this paste after curing.

Tendon — A high strength steel element consisting of one or more wires, strands or bars, or a bundle of such elements, used to impart prestressing forces to the concrete. In post-tensioned applications, a complete assembly consisting of anchorages, prestressing steel (strand), corrosion inhibiting coating and sheathing. It imparts the prestressing force to the concrete.

Tolerance — Specified permissible variation from stated requirements such as dimensions, location, alignment, strength, and air entrainment.

Product tolerances — Those variations in dimensions relating to individual precast concrete members.

Erection tolerances — Those variations in dimensions required for acceptable matching of precast members after they are erected.

Interfacing tolerances — Those variations in dimensions associated with other materials in contact with or in close proximity to precast concrete.

Transfer strength — The minimum concrete strength specified for the individual concrete elements before the prestressing force may be transferred to them, sometimes called detensioning strength or release strength.

Unbonded tendon — A tendon in which the prestressing steel (strand) is prevented from bonding to the concrete. When unbonded tendons are used, prestressing force is permanently transferred to the concrete by the anchorage only.

Veneered construction — The attachment of other materials, such as natural stone or clay products, to a concrete panel.

Wedges — Pieces of tapered metal with teeth which bite into the prestressing steel (strand) during transfer of the prestressing force. The teeth are bevelled to assure gradual development of the tendon force over the length of the wedge.

Wedge set — The relative movement of the wedges into the anchorage cavity during the transfer of the prestressing force to the anchorage.

Workability — The ease with which a given set of materials can be mixed into concrete and subsequently handled, transported, placed, and finished with a minimum loss of homogeneity.

DIVISION 1 - QUALITY SYSTEM

Standard

1.1 Objective

The general objective of this standard is to define the required minimum practices for the production of architectural precast concrete units and for a program of quality control to monitor the production by measurement or comparison to acceptable standards.

Quality control shall be an accepted and functioning part of the plant operation. Overall product quality results from individual as well as corporate efforts. Plant management must make a commitment to quality before it can be effectively adopted or implemented at the operational level. Management shall establish a corporate standard of quality based on uniform practices in all stages of production, and shall require strict observance of such practices by all levels of personnel.

1.2 Plant Quality Assurance Program

1.2.1 General

The plant shall implement and maintain a documented quality assurance program in addition to this manual (MNL-117). Each plant shall have a unique manual based on operations at that facility.

The quality assurance program document shall, as a minimum, cover the following:

a. Management commitment to quality.

Commentary

C1.1 Objective

The individuals in control of operations should have the desire to produce products of proper quality, and should delegate the necessary authority and assign the necessary responsibility for results. Quality products can be expected if proper procedures are established and then carried out.

While the guidelines in this standard deal with the quality control function, it is recognized that the primary responsibility for quality in any product rests with those people who are involved in its production. Production personnel are responsible for quality, and they must understand the role of quality control and work to assure effective monitoring and timely response for necessary correction and improvement.

Supervisory personnel are an integral part of the process and should be committed to quality standards. The production of quality products requires uniformity in expectation by management for all areas of operations and types of products.

Construction project specifications and manuals can prescribe and explain proper quality control criteria for all phases of production consistent with products of the highest quality. To ensure that these criteria are followed, inspection personnel and a regular program of auditing all aspects of production should be provided in all plants.

Although production personnel should be responsible for quality of products, prudence dictates a system of checks and balances which inspection of operations and products provides. Quality control inspections are a necessary and vital management tool to monitor production. This inspection does not remove the responsibilities of the production staff, but provides management with an objective review. The number of persons required to perform services will vary with the size and extent of plant operations.

C1.2 Plant Quality Assurance Program

C1.2.1 General

The use of a written plant practices manual requires an initial effort and commitment by plant management for development of the document and should include an ongoing means of practice guideline development in accordance with the changing nature of products or size of the production facility.

Plant procedures should be documented in the form of specific instructions to operating personnel in a manner that will assure uniformity of both operations and training for present and new employees. See Appendix A for guidelines in developing a

Standard

b. Organizational structure and relationships, responsibilities and qualifications of key personnel.

c. Management review of the quality assurance program at regular intervals, not to exceed one year, to ensure its continuing suitability and effectiveness. This review will include non-conformance, corrective actions and customer complaints.

d. Plant facilities in the form of a plant layout, noting allocation of areas, services, machinery and equipment.

e. Purchasing procedures for quality control compliance, including project specification review for specific requirements.

f. Identification of training needs and provisions for training personnel in quality assurance requirements.

g. Control, calibration and maintenance of necessary inspection, measuring and test apparatus.

h. Uniform methods for reporting, (include sample forms), reviewing and maintaining records. Each precast concrete unit shall be traceable to a specific set of applicable quality control records.

i. Standards for shop (erection and production) drawings to ensure accuracy and uniform interpretation of instructions for manufacturing and handling.

j. Procedures for review and dissemination of project specific requirements to production and quality control personnel.

1.2.2 Documented Procedures

Control of documented procedures and data, relative to the effective functioning of the quality assurance program, shall cover as a minimum:

a. Inspecting and verifying purchased materials for conformance with specification requirements. Vendors shall be required to submit proof of compliance for both materials and workmanship.

b. Sampling methods and frequency of tests.

c. Checking and approval of shop drawings.

d. Inspecting and verifying the accuracy of dimensions. Checking adequacy and sealing of the mold to produce units without undesirable

Commentary

plant quality system manual.

When the listed items are closely controlled through adequate inspection, testing, and documentation review, the objective of quality control — the production at minimal practical cost of products of uniform quality to assure satisfactory service through the intended operating life — will be realized. The best of materials and design practice will not be sufficiently effective unless the actual production practices and procedures are performed properly. Wherever possible, highly varied practices, which are subject to human error or mistakes in judgement among the various production groups, should be eliminated.

The most important aspects of a quality assurance program are:

a. Adequate inspection personnel to assure review of all materials and processes.

b. Clearly defined responsibilities and required function for each inspector.

c. Recognition and acceptance by supervisory personnel of the importance and impact of the quality assurance program and a uniform plant standard of quality.

d. Clear and complete records of inspection and testing.

e. Updating and calibration of testing equipment in a timely manner.

Information gained through quality control inspections should be reviewed on a weekly basis with production personnel. This review may be useful in identifying areas that may require retraining in proper production procedures or modified procedures, or equipment that needs to be repaired or replaced.

C1.2.2 Documented Procedures

A complete and accurate record of operations and inspection thereof is beneficial to a producer if questions are raised during the use of plant products.

See Division 6, *Quality Control* for additional information.

Standard

distortion or surface non-compliances.

e. Procedures for and inspection of batching, mixing, material handling, placing, consolidating, curing, finishing, repairing, product handling, storing and loading.

f. Inspecting the fabrication, placement and securing of reinforcement and hardware, quantity, location and attachment of cast-in items, blockout, and surface features.

g. Inspection of tensioning operations to ensure conformance with specified procedures.

h. Preparing or evaluating mix designs for desired properties.

i. Sampling and testing of materials, including concrete, as required.

j. Inspecting detensioning and stripping procedures.

k. Inspecting all finished products for conformance with shop drawings, approved samples and project requirements.

l. Repair procedures for finish non-compliances.

m. Preparing and maintaining complete quality control records.

n. Maintenance and calibration requirements (items and frequency) of plant equipment affecting product quality.

1.2.3 Management Responsibilities

In order to achieve a satisfactory level of quality control, certain fundamental requirements shall be met by management. These include, but are not limited to:

a. A standard of quality shall be set and observed.

b. Utilization of a written plant manual which will establish a uniform order or practice for all manufacturing operations.

c. Personnel, whose primary function is quality control, shall be responsible to the general manager or chief engineer in functional structure.

d. Establish an acceptance program of finished products prior to shipping.

e. Establish uniform methods for reporting, reviewing and keeping records. Each precast

Commentary

C1.2.3 Management Responsibilities

Plant management must be committed to quality, and this commitment must be demonstrated to all personnel. Quality control inspection functions cannot overcome a lack of dedication to quality by management. Those responsible for producing the product must understand that management is interested in producing a high quality product.

Standard

concrete unit produced shall be traceable to a specific set of applicable quality control records.

f. Establish engineering operations to meet required codes, standards, specifications, and in-plant performance requirements.

1.3 Personnel

1.3.1 General

Each plant shall have personnel qualified to perform the functions of the various positions outlined in this section, and shall clearly define responsibilities and establish the relationship between quality control, engineering, and production.

At least one individual in the plant organization shall be certified as a Level I Technician/Inspector in the PCI Quality Control Personnel Certification program.

1.3.2 Engineering

Plants shall have available the services of a registered professional engineer experienced in the design of precast concrete. The precast engineer shall prescribe design policies for precast concrete elements and be competent to review designs prepared by others. The precast engineer shall be responsible for the design of all products for production, handling, and erection stresses.

The precast engineer shall be responsible for prescribing or approving methods and procedures for tensioning, computations and measurements for elongations, camber and deflections, compensations for operational stress variations and any other variables related to prestressing that may affect the quality of the product.

1.3.3 Drafting

Plants shall utilize experienced personnel competent to prepare shop (production and erection) drawings in general accordance with the "PCI Drafting Handbook — Precast and Prestressed Concrete", MNL-128.

Commentary

C1.3 Personnel

C1.3.1 General

In this section, the functional responsibilities of certain basic positions are outlined. Whether one or more of these functions is assigned to one person, or whether several persons are assigned to a specific function, is the prerogative of management and will depend on the purpose of the product and the size of the plant.

Proper and responsible performance of persons involved in the manufacture of precast concrete products requires specialized technical knowledge and experience.

The PCI Personnel Certification program currently outlines training programs and certification of personnel at three levels. It is recommended that all personnel doing precast concrete inspection and testing work as described in this Manual be certified at the appropriate level. In some operations, the certified individual may perform several tasks, including quality control functions.

C1.3.2 Engineering

Engineering personnel should review the design of precast concrete elements prepared by the engineer of record. The precast engineer should have the ability to solve problems and devise methods, as required, for the design, production, handling and/or erection of precast concrete products. The engineer of record should also approve the sequence of erection, when sequence may affect structural stability of supporting elements which they designed.

C1.3.3 Drafting

Shop drawing details should be of such clarity and completeness to reflect the contract documents in a manner that minimizes the possibility of errors during the manufacturing and erection processes.

Standard

1.3.4 Production

Production personnel shall be qualified to produce units in accordance with the production drawings, approved samples and the plant's quality control requirements.

1.3.5 Quality Control

This function shall have lines of communication to engineering, production and management with responsibility only to management. Quality control personnel shall not report to production personnel.

Responsibilities shall include assuring that the following activities are performed at a frequency shown to be adequate to meet quality objectives or as prescribed in this manual.

a. Inspecting and verifying the accuracy of dimensions and condition of molds.

b. Verifying batching, mixing, material handling, placing, consolidating, curing, product handling and storage procedures.

c. Verifying the proper fabrication and placement of reinforcement, and quantity and location of cast-in items.

d. Inspecting tensioning operations to ensure conformance with specified procedures.

e. Preparing or evaluating mix designs.

f. Taking representative test samples and performing all required testing.

g. Inspecting finished products for conformance with shop drawings, approved samples and project requirements.

h. Preparing and maintaining complete quality control records.

1.4 Design Responsibilities

1.4.1 General

The manufacturer shall be responsible for translating the project requirements into samples, shop drawings, tooling, manufacturing and installation procedures. The manufacturer shall analyze all precast concrete units for anticipated handling stresses or temporary loadings imposed on them prior to and during final incorporation into the finished structure. The manufacturer shall design and provide reinforcement or temporary strengthening

Commentary

C1.3.4 Production

Production personnel have the immediate responsibility of supervising all shop operations involved in the manufacture of products to ensure compliance with production drawings, specifications and established plant standards.

C1.3.5 Quality Control

Quality assurance is the primary responsibility of a plant's quality control staff. Production personnel may be involved in some quality activities but results and records should be audited by quality control personnel for evaluation.

The qualifications of personnel conducting inspections and tests are critical to providing adequate assurance that the precast concrete products will satisfy the desired level of quality.

Quality control personnel should observe and report any changes in plant equipment, working conditions, weather and other items which have the potential for affecting the quality of products.

C1.4 Design Responsibilities

C1.4.1 General

Local practices regarding the design of precast concrete units vary widely as do, to a lesser extent, relevant codes or statutes governing professional design and the responsibility of manufacturers. Hence the points made in this section should be evaluated for conditions applicable to the particular location or to individual projects.

In the interest of both precaster and architect/engineer, the design responsibilities of each party should be clearly defined. It is recommended that this be done in the contract documents.

Standard

of the units to ensure that no adverse stresses are introduced which will exceed the limitations of codes or standards governing the project.

1.4.2 Shop Drawings

The manufacturer shall prepare and submit erection drawings and product details as required for approval in general accordance with the "PCI Drafting Handbook — Precast and Prestressed Concrete", MNL-128, and the project specifications.

Production drawings shall be prepared to convey all pertinent information necessary for fabrication and inspection of the precast concrete products.

The manufacturer shall have a procedure for keeping project documents for at least 5 years following project completion, or as required by local statutes.

1.5 Project Samples

1.5.1 General

After award of the contract, and before producing any units, the precast concrete manufacturer shall prepare and submit for approval, a representative sample or samples of the required color and texture. If the back face of a precast concrete unit is to be exposed, samples of the workmanship, color, and texture of the backing shall be shown as well as the facing. Any changes in the source of materials or in mix proportions to facilitate production shall require new reference samples and approval review. If specified, or if the color or appearance of the cement or the aggregates is likely to vary significantly, samples showing the expected range of variations shall be supplied.

The concrete placement and consolidation method used to make samples shall be representative of the intended production methods.

1.5.2 Size and Shape

Samples shall reflect a shape relationship to the actual casting. Flat samples shall only be used for flat castings.

The size of the samples shall reflect the relationship between materials (maximum size of aggregate to be used), finishes, shapes and casting

Commentary

The responsibilities of the architect/engineer are subject to contractual relationships with the owner and are discussed in Appendix B, Design Responsibilities and Considerations.

C1.4.2 Shop Drawings

The primary function of precast concrete shop (erection and production) drawings is the translation of contract documents into usable information for accurate and efficient manufacture, handling and erection of the units. The erection drawings provide the architect/engineer with a means of checking the interface with adjacent materials and the precaster's interpretation of the contract drawings. Good production drawings provide effective communications between the engineering/drafting and the production/erection departments of a precast concrete plant.

C1.5 Project Samples

C1.5.1 General

All project samples should be submitted promptly for early acceptance to provide sufficient lead time for procurement of materials and production of units for the project.

Sample approval should be in writing with reference to the correct sample code number and written on the sample itself. Approval of the sample by the architect/engineer should indicate authorization to proceed with production, unless such authorization is expressly withheld.

If small samples are used to select the aggregate color, the architect should be made aware that the general appearance of large areas of a building wall may vary from the samples.

Color selection should be made under lighting conditions similar to those under which the precast concrete will be used, such as the strong light and shadows of natural daylight.

C1.5.2 Size and Shape

For non-planar, curved, or other complex shapes, a flat-cast sample may not represent the anticipated appearance of the final product. Select sample shapes that will offer a reasonable comparison to the precast units represented.

Standard

techniques, such as mold types, orientation of exposed surfaces during casting, and consolidation procedures.

1.5.3 Identification

Samples shall be supplied for each of the different finishes for a project and all submitted samples shall be clearly identified.

1.5.4 Visual Mockups and Initial Production Approval of Architectural Finishes

Projects which use new shapes, shape combinations, finishes or aggregates shall utilize a mockup consisting of a full-scale portion of a unit for initial production approval. Previously completed projects, mockups or samples may be used for initial production approval of products which are similar. When approved, these samples shall form the basis of judgment for the purpose of accepting the appearance of finishes. These samples shall establish the range of acceptability with respect to color and texture variations, surface defects, and overall appearance. Samples shall be viewed at a distance consistent with their viewing distance on the structure but not less than 20 ft (6 m). Samples shall also serve as testing areas for remedial work. Approved samples shall be kept at the manufacturing facility and shall be used to monitor the acceptability of the production panels.

The samples shall be stored outdoors and positioned to allow comparison with production units. They shall be stored adjacent to each other to allow proper lighting (sun and shade) for daily comparisons of the production units for finish and exposure.

Commentary

C1.5.4 Visual Mockups and Initial Production Approval of Architectural Finishes

The production of uniform, blemish-free samples, which demonstrate the abilities of a single master craftsman, will be completely misleading and could cause endless difficulties when the production personnel, using actual manufacturing facilities, have to match "the sample". Samples should be made as nearly as possible in the same manner intended for the actual units.

Small 12 in. (300 mm) square samples do not generally reflect the relationship between materials, finishes, shapes, casting techniques, mold types, thickness of concrete section, orientation of exposed surfaces during production and consolidation procedures. Where mockup units are not used, the manufacturer should request the architect/engineer and/or owner to inspect and approve (sign and date) initial production units. Larger production samples will remove uncertainties in the minds of the architect/engineer and owner alike.

At least three range samples of a size sufficient to demonstrate actual planned production conditions may be used to establish a range of acceptability with respect to color and texture variations, uniformity of returns, frequency, size and uniformity of air voids distribution, surface blemishes and overall appearance. When specified, the acceptability of repair techniques for chips, spalls or other surface blemishes should also be established on these samples.

See Appendix C, Finish Samples for additional information.

DIVISION 2 – PRODUCTION PRACTICES

Standard

2.1 General Objectives and Safety

2.1.1 General

The plant facility shall be adequate for production, finish processing, handling and storage of product in accordance with this Manual.

2.1.2 Plant Safety

Each operation shall establish and maintain a written program that encourages workers' safety and health. It shall be patterned after OSHA Safety and Health Standard - 29CFR 1910 and/or other jurisdictional safety and health standards.

2.2 Production and Curing Facilities

2.2.1 Area Requirements

The production and curing areas shall be designed for controlled production of quality architectural precast concrete units and be of adequate size in relation to the volume and characteristics of the products manufactured for a well-organized, continuous operation. Consideration for production flexibility and flow patterns shall be made keeping in mind health and safety provisions.

All materials shall be stored in a manner that will prevent contamination or deterioration and in ac-

Commentary

C2.1 General Objectives and Safety

C2.1.1 General

Plant facilities represent the tools of the industry and as such should be maintained in good operating condition. Facilities suitable for the production of precast concrete units will vary from plant to plant. These facilities will be affected by size, weight, and volume of units produced, variety of surface finishes offered, climate.

C2.1.2 Plant Safety

A safety program is an important element of any production operation. PCI encourages safety and loss prevention programs. The PCI Safety and Loss Prevention Manual, SLP-100, can be used to start a program. Such programs should outline general safety practices as they relate to the precast concrete industry and existing federal regulations.

A safety program should include aspects of the following basic elements:

1. Policy. A written statement of plant or company policy for safety with clear lines of authority should be developed.

2. Rules. The management of each plant should develop a set of safety rules designed to help employees avoid injury.

3. Training. Each plant should have a training program to ensure new and old employees are instructed in safe daily operating procedures.

4. Accident investigation. Accident investigation by management may identify causes or areas needing improvement, better supervision, or employee training.

The details of a safety program are not specified in this Manual, but are left to the individual plants to tailor to their facilities, products and operations.

C2.2 Production and Curing Facilities

C2.2.1 Area Requirements

The molds should be protected against detrimental environmental conditions. The production areas should provide a reasonably uniform ambient environment to maintain the desired concrete temperature. Adequate lighting should be provided for all operations.

Standard

cordance with the manufacturer's instructions, if applicable.

2.2.2 Mold Fabrication

Mold fabrication facilities shall be tooled to provide for the building of molds to a level of accuracy sufficient to maintain the product within required tolerances.

Molds shall be stored in such a manner to protect them from damage that could result in dimensional change or general surface or structural degradation.

For the fabrication of prestressed architectural products, self-stressing molds or forms, bed abutments and anchorages shall be designed by qualified engineers. Information on the capacity of each bed and self-stressing form in terms of allowable prestress force and its corresponding height of application above the form base shall be kept on file.

2.2.3 Storage of Release Agents and Retarders

Release agents and retarders shall be stored in accordance with manufacturer's recommendations, particularly with regard to temperature extremes. Before use, release agents and retarders shall be checked for sediment. If solids have settled out, uniformity and original consistency shall be maintained by periodic mechanical mixing or stirring in accordance with manufacturers recommendations.

Release agents and retarders containing volatile solvents shall be stored in airtight containers to prevent a change in concentration. Release agents shall not be diluted unless specifically permitted by the manufacturer.

2.2.4 Hardware Fabrication and Storage

Materials shall be handled and stored to avoid distortion beyond allowable variations. Steel without corrosion protection shall be stored on pallets, blocks, racks or in containers as well as protected from contamination.

Stainless steel hardware shall be protected from contamination from other metals during storage and fabrication. Stainless steel hardware shall be handled with nonmetallic or stainless materials only.

Adequate space and equipment shall be provided for the fabrication of hardware. Fabrication equipment for hardware shall be of a type, capacity, and

Commentary

C2.2.2 Mold Fabrication

Mold fabrication facilities should be capable of maintaining constant working temperatures above the minimum required for specific raw materials and processes used. Resins, catalysts, accelerators and acetone, etc., for plastic molds should be stored within the manufacturer's recommended temperature range and away from the production areas.

The design of prestressing forms and beds should be based upon stated factors of safety according to sound engineering principles taking into account the magnitude, position and frequency of the forces anticipated to be impacted onto the form or bed. Foundations should be sufficiently firm to prevent undesirable movements.

C2.2.3 Storage of Release Agents and Retarders

Release agents and retarders should have a reasonably long and stable storage life, but may be susceptible to damage from extreme temperature changes.

Certain weather conditions can affect the performance of water-based release agents or retarders. Generally, they can't be used in cold weather because they might freeze. Even at temperatures slightly above freezing, some water-based products thicken enough to produce more bugholes and/or reduce performance.

Some oils have a critical emulsifier content and dilution makes the emulsion unstable and causes poor performance.

Standard

accuracy capable of fabricating hardware assemblies to the required tolerances and quality.

Electrodes used for welding operations shall be bought in hermetically sealed containers. Low hydrogen SMAW electrodes shall be stored either in their original air tight containers or at the recommended elevated temperature in a suitable oven. Once containers are opened, welding electrodes and wires shall be kept in dry heated storage.

If hardware is fabricated by an outside supplier, that supplier shall furnish records of compliance to specification requirements and mill certificates for material used.

Periodic review of hardware fabrication shall be performed by quality control personnel at the fabrication area.

2.2.5 Casting Area and Equipment

The concrete handling equipment shall be such that it will convey concrete from the mixer to the mold:

1. In sufficient quantities to avoid undue delays in placement and consolidation.
2. Without segregation of aggregates and paste.
3. With uniform consistency.
4. With ease of discharge into molds.
5. With equipment capable of being thoroughly cleaned.
6. With consideration for concrete temperatures.

The casting area shall be supplied with equipment in good operating condition for consolidation of the concrete after placement in the mold. Concrete vibrators may be internal, external or surface types, depending on the service required. These vibrators may be used singly or in various combinations. Provisions shall be made to supply adequate and safe power for these units.

Before a vibratory unit is put into use, it shall be

Commentary

It is necessary to store electrodes in a controlled environment to prevent moisture absorption into the flux from ambient humidity.

C2.2.5 Casting Area and Equipment

The casting area should provide flexibility in planning and spacing of the molds and efficient movement of workers, materials and equipment involved in placing, initial curing and stripping of products. Facilities should be provided for post-tensioning or pretensioning, if required.

Provision should be made to control the temperature with reasonable accuracy, since chemical retarders and form release agents may react differently under varying temperature and humidity ranges. To meet the provisions of Division 4, it may be necessary to provide heating and/or proper ventilation for the casting area and the molds, depending on the geographic location of the plant.

Standard

checked to verify that it is working properly.

2.2.6 Curing and Finishing Areas

The plant shall be capable of maintaining a minimum concrete temperature of 50 deg. F (10 deg. C) during the initial curing cycle (prior to stripping).

When moist curing is used, facilities shall provide a well drained area with adequate covering to maintain the required relative humidity and temperature.

The capacity of the heat source for accelerated curing shall be related to the volume of concrete to be cured, the stripping or transfer strength level, the length of the curing cycle, and the effectiveness of the heat enclosure. The heat source and the distribution system shall be protected from operational hazards and shall provide uniform controlled heat for each unit or series of units being cured. Heat enclosures shall not damage the products nor shall they affect the uniformity of heat distribution to the units. Equipment shall be available to control or record the time and temperature relationship for the accelerated curing cycle. The number of thermometers or thermocouples shall be sufficient to establish that uniform heat is supplied to each unit.

The finishing areas shall provide for the varying types of finishes to be produced by the plant (see Article 2.7).

In the case where units are to receive no further finish treatment prior to storage and delivery, they shall be protected from damage that cannot be readily removed by cleaning. Products that are to receive a sealer shall be treated and cured as required by the sealer manufacturer.

Provision shall be made for an area to patch flawed or damaged products.

2.2.7 Handling Equipment

The production facilities shall include adequate product handling equipment maintained in good working condition. Handling equipment shall be capable of stripping, moving, stacking, retrieving, and loading units without damaging the products.

Commentary

C2.2.6 Curing and Finishing Areas

Configuration of curing facilities will depend on the components being made and the method of curing.

One temperature measuring device should be used for each set of units cast: (1) within one hour of each other; (2) with similar mix; and (3) cured under the same methods/controls.

Temperature measuring devices should be located near the concrete test units, if they are cured with the product, or at a location furthest from the heat source, if a master-slave control is used.

Lighting is extremely important in the finishing area and at the point where final inspection is made before transport to the storage area. This is where comparison to the approved samples is made for color and texture. Where possible, indoor lighting should compare to daylight as closely as possible.

C2.2.7 Handling Equipment

The type and capacity of equipment for handling finished products will depend on the actual products and the operating conditions.

Standard

2.2.8 Storage Area for Finished Products

The storage area shall be of adequate size to allow safe storage and easy access to the products by the handling equipment. The area shall be clean, well-drained and stabilized to minimize differential settlements under all weather conditions and to minimize soiling, warping, bowing, chipping or cracking of the product.

Storage racks shall be designed to safely store product to prevent units from tipping over and damaging adjacent units. Storage racks such as horses, A-frames, and vertical racks shall be well constructed to minimize warping, bowing, chipping, or cracking of the products. Storage systems actually carrying products shall be reviewed by the plant engineer for their safe load capacity. Where necessary, such storage equipment shall be protectively coated to avoid any staining or discoloration of the finished products.

2.3 Welding

2.3.1 Structural Steel

Welding of steel plates, angles, and other shapes shall be in conformance with AWS D1.1. All welds shall be performed by welders qualified in accordance with AWS D1.1 and D1.4. Welding procedure specifications for structural steel shall be written by plant engineering or quality control. Welding current shall be within the range recommended by the electrode manufacturer (typically on the side of the package). The size, length, type, and location of all welds shall conform to those shown on the shop drawings, and no unspecified welds shall be added without approval of the precast engineer. Surfaces to be welded, and surfaces adjacent to a weld, shall be uniform, free from fins, tears, cracks and other discontinuites and free from loose or thick scale, slag, rust, moisture, grease, and other foreign material that would prevent proper welding or produce objectionable fumes. Mill scale that can withstand vigorous wire brushing, a thin rust-inhibitive coating, or antispatter compound may remain.

The preheat requirements of insert plates shall be in accordance with Table 2.3.1. Welding shall not be done when the temperature in the immediate vicinity of the weld is below 0 deg. F (-18 deg. C). When the insert plate temperature is below 32 deg. F (0 deg. C), the plate shall be preheated to a temperature of at least 70 deg. F (21 deg. C) and this

Commentary

C2.2.8 Storage Area for Finished Products

Access should be provided in the storage facilities to allow for product inspection.

The subgrade in the storage area should be stabilized to avoid soft spots where one end of a member can settle. This settlement creates twisting or tensile stresses which can cause cracking and damage. For example, units should not be stored on frozen ground without proper safeguards to prevent settlement when the ground thaws. The storage area should be inspected after hard rains or large snow melts for washouts and other damage.

C2.3 Welding

C2.3.1 Structural Steel

Temperature sensitive crayons are frequently used to give an approximate preheat temperature indication. This measurement usually should be made within one inch (25 mm) of the weld on the base metal. Crayon marks should never be made directly on the weld because of possible contamination.

Standard

Table 2.3.1. Minimum preheat temperatures for insert plates[1,2]

Thickest Section at Point of Welding, in. (mm) for A36, A441, A500, Grades A and B, and A572, Grades 42 and 50[2]	Minimum Temperature [1][3] °F (°C)
Up to 3/4 (19)	None
Over 3/4 through 1-1/2 (19-38.1 incl)	50 (10)
Over 1-1/2 through 2-1/2 (38.1-63.5 incl)	150 (66)
Over 2-1/2 (63.5)	225 (107)

(1) Minimum temperature required when using shielded metal arc welding with low hydrogen electrodes, submerged arc welding, gas metal arc welding or flux cored arc welding.

(2) If the steel specification for the insert plate or welding process being used is not shown, refer to Table 4.3 of AWS D1.1 for preheat requirements.

(3) This must be compared to the preheat requirements for the reinforcing bar. The higher of the two requirements should be applied.

temperature maintained throughout the entire welding process. Three inches of the insert plate, in each direction around the weld, shall be preheated to at least the minimum required temperature.

Slag from each pass shall be completely removed before depositing the next pass to avoid porosity and slag entrapment. Slag shall be removed from all completed welds, and the weld and adjacent base metal shall be cleaned by brushing or other suitable means. Tightly adherent spatter remaining after the cleaning operation is acceptable. Accessible welds of corrosion protected material (galvanized or painted) shall be touched up after welding. Zinc-rich paint shall be brush or spray applied to a thickness of approximately 0.004 in. (100 μm) over the welded areas to replace the removed galvanizing or in conformance with ASTM A780.

One of the two following procedures shall be used for welding galvanized steel:

1. Removal of galvanizing on mating surfaces to be welded and use of standard procedures.

2. Use of welding procedures using galvanized base metal and qualified by test in accordance with the AWS D1.1.

Commentary

When galvanized steel is welded, some of the zinc coating is volatilized on each side of the weld and, while a thin layer of zinc-iron alloy remains, there is a loss in corrosion resistance. In the case of zinc-rich painted steel, welding causes decomposition of the paint film which is burnt off for some distance each side of the weld. The width of the damaged zone will depend on the heat input.

Zinc-rich paints are available which have been specially formulated such that it is not necessary to remove the coating from the weld path prior to welding. A letter should be obtained from paint manufacturer stating it to be "weldable."

Some galvanizers supply steel with predetermined areas uncoated to enable the steel to be welded without concern for problems that can arise from the presence of zinc during welding.

Standard

When welding stainless steel plates to other stainless steel or to low carbon steel, the general procedure for welding low carbon steels shall be followed, taking into account the stainless steel characteristics that differ, such as higher thermal expansion and lower thermal conductivity. Welding of stainless steel shall be in conformance with AWS B2.1, AWS A5.4 and AWS D1.1. Welding of stainless steel shall be done by qualified welders familiar with the welding requirements of these alloys. Preheating or postheating is not necessary.

Commentary

It is not uncommon to find small cracks in a fillet weld on galvanized steel, extending from the root toward the face of the bead. Whether cracking will occur depends on many factors such as the silicon content of the weld metal, the degree of penetration of the weld, the gap between metals, the thickness of the base metal which influences restraint of the joint, and the coating weight of the zinc and the microstructure of the zinc coating which are both influenced by the composition of the base plate, particularly with respect to silicon content. Low-silicon or rutile (non-low-hydrogen) base electrodes with low-silicon content (0.2Si or lower) generally reduce cracking.

Removal of zinc, where weld is to be placed, is the most conservative approach in welding galvanized steel. Welding procedures will then be the same as for uncoated steel. Zinc can be removed by burning with an oxygen fuel gas torch, by shot blasting with portable equipment or grinding with abrasive discs.

In general, manual metal arc welding procedures for galvanized steel are similar to welding uncoated steel. However, welding of galvanized steel generally requires that the welder receive specialized training. In addition, qualification of the welder and welding procedure using the thickest coating anticipated is strongly recommended.

Stainless steels have many properties that differ from those of carbon or other steels and the differences become more pronounced with higher chromium contents. For example:

1. Their thermal conductivity is much lower so they are more susceptible to local overheating and to distortion when they are welded.

2. Their thermal expansion is higher and this tends to increase distortion and results in higher stresses on the weld during cooling.

3. They resist oxidation until heated to temperatures around their melting points in the presence of air. Then a highly refractory chromium oxide is formed preventing these alloys from being cut with an ordinary oxyacetylene cutting torch. In welding stainless steels, the molten metal must be well protected from the air.

4. Some martensitic alloys, as the carbon content increases, are highly hardenable and become brittle when heated and cooled, due to excessive grain growth at high temperatures. Other martensitic alloys suffer a loss in corrosion resistance if there is appreciable carbon in the base or weld metal.

Because of the relatively high coefficient of thermal expansion [9.2 μ in./in. deg. F (16.6 μm/m deg. C)] and lower thermal conductivity [9.1 Btu/hr ft deg. F (15.7 w/mk)] of austenitic stainless steel, precautions are necessary to avoid weld bead cracking, minimize distortion of steel and avoid cracking of concrete. The following procedures may be used to minimize these problems: lower weld current settings, skip-weld techniques to minimize heat concentration, use of copper backup chill bars or other cooling techniques to dissipate heat, tack welding to hold the parts in alignment during welding and small weld passes.

Stainless steels cannot be cut smooth with an ordinary oxy-

Standard

A method for cutting stainless steel shall be used that produces a clean, smooth edge. Powder cutting processes such as plasma arc shall be used.

The edges of a thermally cut weld joint shall be cleaned by machining or grinding to remove surface contamination, particularly iron. Surfaces to be welded shall be sanded smooth, not ground, and all blue heat tint removed. Parts to be joined shall also be free of oil, grease, paint, dirt, and other contaminants.

In joining austenitic stainless steels to carbon steels or low-alloy steels, a stainless steel welding rod that is sufficiently high in total alloy content, such as Type 309, shall be used. When, due to service requirements, the depositing of carbon steel or low-alloy steel weld metal on stainless steel is required, the short-circuiting method of metal transfer shall be used.

Welds and the surrounding area on stainless steel shall be cleaned of weld spatter, flux, or scale to avoid impairment of corrosion resistance.

2.3.2 Reinforcement

Welding procedure specifications for reinforcing bars shall be written by plant engineering or quality control. Welding of reinforcing bars shall be executed considering steel weldability and proper welding procedures, whether performed in-plant or by an outside supplier.

Commentary

acetylene torch.

Cleaning of stainless steel to be welded is important. Contamination from grease and oil can lead to carburization in the weld area with subsequent reduction of corrosion resistance. Post weld clean-up is also important and should not be done with carbon steel files and brushes. Stainless steel wire brushes should be used. Carbon steel cleaning tools, as well as grinding wheels that are used on carbon steel, can leave fine articles embedded in the stainless steel surface that will later rust and stain if not removed by chemical cleaning.

This will prevent martensite formation while at the same time preserving residual amounts of ferrite, which counteract the tendencies for hot cracking (at the time of welding) even under conditions of severe restraint.

C2.3.2 Reinforcement

The grade, bar size and chemical composition of the reinforcing bar should be known prior to welding. Assuming a carbon equivalent higher than what actually exists could result in a more complicated and costly welding process. The weldability of steel established by its chemical analysis limits the applicable welding procedures and sets preheat and interpass temperature requirements. Weldability properties are specifically excluded from the ASTM A615, A616, and A617 specifications. Most reinforcing bars which meet ASTM A615, Grade 60 (420), will require preheating. A615, Grade 40 (300) bars may or may not require preheating. ASTM A706 bars are specially formulated to be weldable; hence, the specification contains chemical composition requirements and calculation of the carbon equivalent.

The ASTM specifications for billet-steel, rail-steel, axle-steel and low-alloy steel reinforcing bars (A 615, A 616, A 617, and A 706 respectively) require identification marks to be rolled into the surface of one side of the bar to denote the producer's mill designation, bar size, type of steel and minimum yield designation, Fig. C2.3.1.

Minimum yield designation is used for Grade 60 (420) and Grade 75 (520) bars only. Grade 60 (420) bars can either have one (1) single longitudinal line (grade line) or the number 60 (420) (grade mark). Grade 75 (520) bars can either have two (2) grade lines or the grade mark 75 (520).

A grade line is smaller and between the two main ribs which are on opposite sides of all U.S. made bars. A grade line must be continued at least 5 deformation spaces. A grade mark is the 4th mark on a bar.

Standard

Welding procedures shall be in conformance with AWS D1.4 using shielded metal arc (SMAW), gas metal arc (GMAW), or flux cored arc (FCAW) processes. Striking an arc on the reinforcing bar outside of the weld area shall not be permitted. Quality control or engineering shall review the mill test report to determine the carbon equivalent (C.E.) and the preheat requirements. Minimum preheat and interpass temperatures for welding of reinforcing bars shall be in accordance with Table 2.3.2 using the highest carbon equivalent number of the base metal. Temperature sensitive crayons shall be used to determine approximate preheat and interpass temperatures.

For billet-steel bars, conforming to ASTM A615/A615M, the carbon equivalent shall be calculated using the chemical composition, as shown in the mill test report, by the following formula:

$$C.E. = \%C + \%Mn/6$$

If mill test reports are not available, chemical analysis may be made on bars representative of the bars to be welded. If the chemical composition is not known or obtained:

1. For bars No. 6 (19) or less, use a minimum preheat of 300 deg. F (150 deg. C).

2. For bars No. 7 (22) or larger, use a minimum preheat of 400 deg. F (200 deg. C).

3. For all ASTM A706 bar sizes, use Table 2.3.2 C.E. values of 0.46 percent to 0.55 percent inclusive.

Surfaces of reinforcing bars to be welded and surfaces adjacent to a weld shall be free from loose or thick scale, slag, rust, moisture, grease, epoxy coating, or other foreign material that would prevent proper welding or produce objectionable fumes. Mill scale that withstands vigorous wire brushing, a thin rust inhibitive coating, or anti-spatter compound, may remain.

The ends of reinforcing bars in direct butt joints shall be shaped to form the weld groove by oxygen

Commentary

Grade 40 (300) and 50 (350) bars are required to have only the first three identification marks (no minimum yield designation).

VARIATIONS: Bar identification marks may be oriented as illustrated or rotated 90°. Grade mark numbers may be placed within separate consecutive deformation spaces. Grade line may be placed on the side opposite the bar marks.

AWS D1.4 indicates that most reinforcing bars can be welded. However, stringent preheat and other quality control measures are required for bars with high carbon equivalents. Except for welding shops with proven quality control procedures that meet AWS D1.4, it is recommended that carbon equivalents be less than 0.45 percent for No. 7 (22) and larger bars, and 0.55 percent for No. 6 (19) and smaller bars.

Welding procedures are critical for reinforcing steel because this steel has a relatively high carbon content. The more carbon the steel contains, the more brittle the material and the higher the susceptibility to embrittlement occurring after a weld begins to cool. As a weld cools, a very brittle form of iron called martensite forms just outside the weld zone. This material is subject to fracture upon impact. For example, if a welded assembly which had not been welded using proper procedures is raised to shoulder height and dropped to the floor, it is quite possible the bars would literally break off at the weld point like a shattered piece of crystal. This is due to the brittle formation of martensite.

Standard

cutting, air carbon arc cutting, sawing, or other mechanical means. Bars for direct butt joints that have sheared ends shall be trimmed back beyond the area deformed by shearing.

Welding shall not be done when ambient temperature is lower than 0 deg. F (-18 deg. C), when surfaces to be welded are exposed to rain, snow, or wind velocities greater than five miles per hour (eight kilometers per hour), or when welders are exposed to inclement conditions.

Table 2.3.2 Welding reinforcing bars

Carbon equivalent[4] range, percent	Size of reinforcing bar	Minimum preheat and interpass temperatures[1,2,3,6] deg F.	deg C.
0.40 max.	Up to 11 (36) inclusive 14 (43) and 18 (57)	None[5] 50	10
0.41-0.45 inclusive	Up to 11 (36) inclusive 14 (43) and 18 (57)	None[5] 100	40
0.46-0.55 inclusive	Up to 6 (19) inclusive 7 (22) to 11 (36) inclusive 14 (43) and 18 (57)	None[5] 50 200	10 90
0.56-0.65 inclusive	Up to 6 (19) inclusive 7 (22) to 11 (36) inclusive 14 (43) and 18 (57)	100 200 300	40 90 150
Above 0.66	Up to 6 (19) inclusive 7 (22) to 18 (57) inclusive	300 400	150 200

Notes:

1. When reinforcing steel is to be welded to main structural steel (insert plates or angles), the preheat requirements of the structural steel shall also be considered (see Table 2.3.1). The minimum preheat requirement to apply in this situation shall be the higher requirement of the two tables. However, extreme caution shall be exercised in the case of welding reinforcing steel to quenched and tempered steels, and such measures shall be taken as to satisfy the preheat requirements for both. If not possible, welding shall not be used to join the two base metals.

2. When the base metal is below the temperature listed for the welding process being used and the size and carbon equivalent range of the bar being welded, it shall be preheated (except as otherwise provided) in such a manner that the cross section of the bar for not less than 6 in. (150 mm) on each side of the joint shall be at or above the desired minimum temperature. If multiple passes are required to make the weld, the area shall be reheated during the time between passes (interpass). Preheat and interpass temperatures shall be sufficient to prevent crack formation.

3. After welding is complete, bars shall be allowed to cool naturally to ambient temperature. Accelerated cooling is prohibited.

4. Where it is impractical to obtain chemical analysis, the carbon equivalent shall be assumed to be above 0.66 percent except for ASTM A706 bars.

5. When the base metal is below 32°F (0°C), the base metal shall be preheated to at least 70°F (21°C) and maintained at this minimum temperature during welding.

6. Use temperature sensitive crayons for determining approximate preheat and interpass temperatures.

Commentary

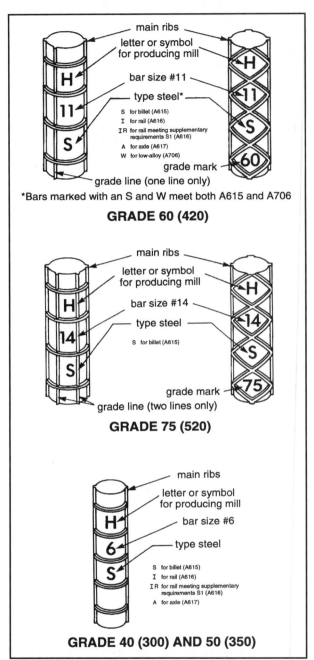

Fig. C2.3.1 Standard identification markings for reinforcing bars.

Metric Bar Size	Diameter (mm)	Inch-Pound Bar Size	Diameter (in.)
#10	9.5	#3	0.375
#13	12.7	#4	0.500
#16	15.9	#5	0.625
#19	19.1	#6	0.750
#22	22.2	#7	0.875
#25	25.4	#8	1.000
#29	28.7	#9	1.128
#32	32.3	#10	1.270
#36	35.8	#11	1.410
#43	43.0	#14	1.693
#57	57.3	#18	2.257

Standard

Preparation for welding on coated base metal shall preferably be made before coating. Welding galvanized metal, without prior removal of the coating, shall be performed in accordance with AWS D1.4 or AWS WZC (D19.0). Welding of galvanized metal may also be done after removing all coating from within 2 in. (50 mm) of the weld joint. The welding shall be performed in accordance with AWS D1.4 for uncoated reinforcing bars. The galvanized coating shall be removed with oxyfuel gas flame, abrasive shot blasting, or other suitable means.

When welding or preheating epoxy coated base metal, the epoxy coating shall be removed from the surfaces to be heated.

After welding, suitable coating protection (zinc-rich or epoxy paint) shall be applied to the finished joint to restore the corrosion resistant properties of the coated bars.

Tack welds that do not become a part of permanent welds shall not be made unless shown on the approved shop drawings — for all reinforcing bars including ASTM A706 bars. Tack welds shall be made in conformance with all the requirements of AWS D1.4, i.e., preheating, slow cooling, use of proper electrodes, and the same quality requirements as permanent welds.

Tack welding shall be carried out without significantly diminishing the effective steel area or the bar area shall be one-third larger than required. A low heat setting shall be used to reduce the undercutting of the effective steel area of the reinforcing bar.

Reinforcing bars which cross shall not be welded unless shown on the approved shop drawings.

Reinforcing bars shall not be welded within 2 bar diameters of the beginning point of tangency of a cold bend.

Commentary

Tack welding, unless done in conformance with AWS D1.4, may produce crystallization (embrittlement or metallurgical notch) of the reinforcing steel in the area of the tack weld. Tack welding seems to be particularly detrimental to ductility (impact resistance) and fatigue resistance, and, to a lesser extent, to static yield strength and ultimate strength. Where a small bar is tack welded to a larger bar, a detrimental "metallurgical notch" effect is exaggerated in the large bar. Fast cooling under cold weather conditions is likely to aggravate these effects.

There is essentially no difference in the potential for embrittlement and chemical change the material undergoes during tack welding as compared to a larger fillet, flare, or flare bevel weld. This is because tack welding produces the same temperatures as structural welding. Therefore, if the reinforcing steel requires the use of preheat and low hydrogen electrodes, it must also be employed when tack welding is used to hold a cage together. Tack welding should be limited to straight sections of bars and within the limitations indicated in the shop drawing section.

When bars are bent cold without the addition of heat, they become sensitive to heat. Subsequently, the application of too much heat will cause the bars to crystallize and result in unpredictable behavior of the reinforcing bar at the bend. Therefore, as a precaution it is necessary to keep welds away from cold bends. While AWS D1.4 suggests allowing a cold bend at two bar diameters from a weld, experience shows that a minimum distance of 2 in. (50 mm) with 3 in. (75 mm) preferred is better with the small bars commonly used in precasting, see Fig. C2.3.2.

Standard

When reinforcing steel is welded to structural steel members, the provisions of AWS D1.1 shall apply to the structural steel component. When joining different grades of steels, the filler metal shall be selected for the lower strength base metal.

2.3.3 Stud Welding

Headed studs and deformed bar anchors used for anchorage shall be welded in accordance with AWS D1.1 and AWS C5.4 The studs and base metal area to be welded shall be free from rust, rust pits, scale, oil, moisture, or other deleterious materials that would adversely affect the welding operation. The base metal shall not be painted, galvanized or cadmium-plated prior to welding. Thickness of plates to which studs are attached shall be at least 1/2 of the diameter of the stud.

The arc shields or ferrules shall be kept dry. Any arc shields which show signs of surface moisture from dew or rain shall be oven dried at 250 deg. F (120 deg. C) for two hours before use. After welding, arc shields shall be broken free from studs. The completed weld shall have a uniform cross section for the full circumference of the stud.

Stud welding equipment settings shall be based on written welding procedure specifications, past practice, or recommendations of the stud and equipment manufacturer. If two or more stud welding guns are to be operated from the same power source, they shall be interlocked so that only one

Commentary

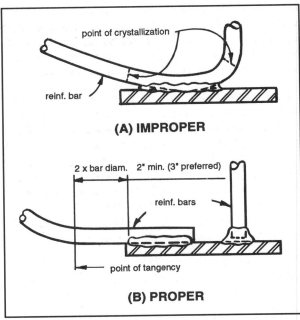

Fig. C2.3.2. Improper and proper methods of welding reinforcing bars.

C2.3.3 Stud Welding

The minimum plate thickness requirement is necessary to prevent melt-through occurring and to assure that the weld is as strong as the stud to utilize its maximum load capacity.

Standard

gun can operate at a time, so that the power source has fully recovered from making one weld before another weld is started. While in operation, the welding gun shall be held in position without movement until the weld metal has solidified.

Stud welding shall not be done when the base metal temperature is below 0 deg. F (-18 deg. C) or when the surface is wet or exposed to falling rain or snow. When the temperature of the base metal is below 32 deg. F (0 deg. C), one additional stud in each 100 studs welded shall be tested by methods specified in Article 6.1.3(9), except that the angle of testing shall be approximately 15 deg. This is in addition to the first two studs tested for each start of a new production period or change in set-up.

When studs are welded using prequalified FCAW, GMAW or SMAW processes, the following requirements shall be met:

1. Surfaces to be welded and surfaces adjacent to a weld shall be free from loose or thick scale, slag, rust, moisture, grease, and other foreign material that would prevent proper welding or produce objectionable fumes.

2. For fillet welds, the end of the stud shall be clean and the stud base prepared so that the base of the stud fits against the base metal. The minimum size of the fillet weld shall be the larger of those required in Table 2.2 or Table 7.2 of AWS D1.1.

3. The base metal to which studs are welded shall be preheated in accordance with the requirements of Table 2.3.1.

4. SMAW welding shall be performed using low hydrogen electrodes 5/32 or 3/16 in. (4 or 5 mm) in diameter, except that a smaller diameter electrode may be used on studs 7/16 in. (11 mm) or less in diameter for out-of-position welds.

2.4 Molds

2.4.1 Materials and Construction

All molds, regardless of material, shall conform to the profiles, dimensions and tolerances indicated by the contract documents and the approved shop drawings.

Molds shall be dimensionally stable to produce the required finish and tolerance. Repeated use of molds shall not affect the dimensions or planes of

Commentary

The stud welding arc is normally of sufficient intensity and duration to burn or vaporize thin layers such as light coatings of paint, scale, rust, or oil. However, initial metal-to-metal contact should be made between the stud and work to draw the arc. Most controllers provide a high-voltage pilot surge to start the arc. A center punch or other mechanical means can be used to penetrate a thick coating.

Low-carbon (mild) steels can be stud welded with no major metallurgical problems. The upper carbon limit for base plate steel to be arc stud welded without preheat is 0.30%.

If medium- and high-carbon base-plate materials are to be stud welded, it is imperative that preheat be used to prevent cracking in the heat-affected zones. In some instances, a combination of preheating and postheating after welding is recommended.

Generally, the high-strength low alloy steels are satisfactorily stud welded when their carbon content is 0.15% or lower. If the carbon content exceeds 0.15%, it may be necessary to preheat the work to a low preheat temperature to obtain desired toughness in the weld area.

Most classes of stainless steel may be stud welded. Only the austenitic stainless steels (300 grades except 303) are recommended for general applications.

C2.4 Molds

C2.4.1 Materials and Construction

The overall quality of the project starts with molds capable of allowing the production personnel to make units which meet all the specifications with ease and accuracy.

The appearance of the finished surface is directly related to the choice of mold material and the quality of the mold itself. The in-service life of a mold is also a function of the choice of

Standard

the molds beyond allowable tolerances. The mold fabrication and set-up tolerances shall be sufficient to produce units within specific tolerances. Mold materials shall not warp or buckle due to temperature change or moisture, which can cause unsightly depressions and uneven swells in the finished surface. The mold materials shall be nonabsorbent or sealed to prevent excessive moisture absorption in order to minimize variations in finish due to differential moisture movements resulting from varying degrees of absorbency. The precaster shall evaluate the effect of different materials on the color of the finished surface when they are combined in the same mold.

Molds shall be coated with release agents that will permit release without damaging or staining the concrete, and without affecting subsequent coating, painting or caulking operations. Release agents shall be applied in accordance with manufacturer's directions. Just prior to applying a release agent, the surfaces of the mold shall be clean and free of water, dust, dirt, or residues that could be transferred to the surface of the concrete or affect the ability of the release agent to function properly. Excess release agent shall be removed from the mold surface prior to casting. Prestressing tendons, anchorages for miscellaneous connections and reinforcement shall not be contaminated by form release agents.

Molds shall be built sufficiently rigid to provide dimensional stability during handling. The assembled mold shall not allow leakage of water or cement paste. Joints in the mold material shall be made so that they will not be reflected in the concrete surface

Commentary

mold material, which must therefore be selected with care. Molds for architectural precast concrete can be made of various materials such as plywood, concrete, steel, plastics, polyester resins reinforced with glass fibers, plaster, or a combination of these materials. For complicated details, molds of plaster, elastomeric rubber, foam plastic, or sculptured sand may be used. These molds are often combined or reinforced with wood or steel depending on the size and complexity of the unit to be produced.

Mold design should take into account the special requirements of precast concrete products. Sharp angles and thin projections should be avoided whenever possible, and chamfers or radii at inside corners of the mold should be incorporated due to the possibility of chipping and spalling at the corners during stripping.

In general, the mold fabrication and set-up tolerances should be one-half of the tolerances of the units they are to produce or a maximum of ±1/8 in. (±3mm).

The selection of a release agent should include investigation of the following factors:

1. Compatibility of the agent with the mold material, form sealer or admixtures in the concrete mix. Because of the rapid loss of slump, most superplasticized concrete needs a smooth "frictionless" surface along which the concrete can easily move.

2. Possible interference with the later application of sealants, sealers, or paints to the mold contact area.

3. Discoloration and staining of the concrete face.

4. Amount of time allowed between application and concrete placement and the minimum and maximum time limits for molds to stay in place before stripping. Release agent may require a curing period before being used. If too fresh, some of the release agent will become embedded in the concrete.

5. Effect of weather and curing conditions on ease of stripping.

6. Uniformity of performance.

7. Meet current local environmental regulations regarding the use of a volatile organic compound (VOC) compliant form release agent.

Applying too much of the release agent can cause excessive surface dusting on the finished concrete.

Mineral oil, oil-solvent based release agents, or paraffin wax should not be used on rubber or elastomeric liners as the hydrocarbon solvent will soften the rubber. The rubber or elastomeric supplier's recommendations should be carefully followed.

Mold seams as well as welds on steel molds should be checked daily during the mold preparation stage. Mold seams resulting from jointing of loose parts, such as bulkheads, side or top molds, or any mold modification pieces should be minimized to the extent required by the finished surface, well-fitted, secured, and sealed to prevent leakage.

Standard

and mar the appearance of the product.

Molds shall be capable of supporting their own weight and the pressure of the fresh concrete without deflection or deformation which exceeds tolerances. Molds shall be sufficiently rigid to withstand the forces necessary for consolidating the concrete. Molds subjected to external vibration shall be capable of transmitting the vibration over a sufficient area in a relatively uniform manner without flexing or plate flutter. The molds shall be designed to ensure that resonant vibrations which may be imparted into local areas of the forms are minimized.

Molds shall permit controlled, fixed positioning or jigging of hardware and allow for the suspension or placement of the reinforcing cage in a position that maintains the specified concrete cover. Blockouts shall be of the size, shape and located as shown on the shop drawings. Blockouts shall be held rigidly in place within tolerance. Mold parts shall allow for stripping without damage to the units.

Molds shall be designed to prevent damage to the concrete from: (1) restraint as the concrete shrinks; (2) the stripping operation; and (3) dimensional changes due to pretensioning.

Wood molds shall be sealed with suitable materials to prevent absorption. The sealer manufacturer's instructions regarding application shall be followed. Surface condition, joints and coating material shall be visually inspected prior to each use.

Concrete molds shall be treated with a coating which renders the concrete non-absorbent to reduce mold damage and to improve the release of the product during the stripping operations.

Plastic molds shall not be used when temperatures above 140 deg. F (60 deg. C) are anticipated. The susceptibility of the plastic mold to attack by the proposed release agents shall be determined prior to usage. Surface conditions, joints and gel coat material shall be visually inspected prior to each use.

Steel molds shall be visually inspected prior to each use for rust, welding distortion and tightness of steel sheet joints. If it is planned to apply a prestressing force by jacking against the form, the mold shall be sufficiently strong to withstand the force without buckling or wrinkling and still maintain the required dimensional tolerances.

Commentary

Molds should have high rigidity, minimum deflection, and minimum movement of the mold material between the stiffeners.

Sealing wood molds minimizes non-uniformity in surface finish and will stabilize the mold dimensions. The manufacturer's instructions regarding application of the sealer should be followed. For some sealers there are minimum temperatures stated below which they must not be applied. An appropriate drying or curing time should be allowed.

Steel molds should be well braced and frequently examined for bulging or buckling. Dimpling, twisting, or bending may occur if they are not properly stacked for storage. When joining two or more steel sheets by welding, care is required to avoid distortion from the heat of the welding operation. If joining is required, the welds should be ground smooth and coated with an epoxy or similar material to hide the joint. Forms should allow for shortening and movements of the precast concrete units during transfer of prestress and application of heat when accelerated curing is used.

Standard

2.4.2 Verification and Maintenance

The mold surfaces and dimensions shall be checked in detail after construction and before the first unit is made and after any modifications. A complete check of the first product from the mold shall be performed. Molds shall be thoroughly cleaned and inspected before each use for defects that will affect performance or appearance.

Bulkheads, templates, and similar equipment shall be regularly inspected and maintained as necessary. All anchorage locations on the mold for holding any cast-in materials to a given position shall be checked for wear. If more than one mold is used to produce a given unit, a comparative dimensional check shall be made.

2.5 Hardware Installation

All connection hardware, anchors, inserts, plates, angles, handling and lifting devices, and other accessories shall be checked prior to casting to verify that they are accounted for, of the proper size and type, and accurately located as detailed on the shop drawings.

Hardware shall be firmly held in the correct position and alignment, during placement, consolidation and finishing of the concrete, by attachment to the reinforcing cage or by jigs, positioning fixtures, mold brackets, or stiffbacks. Hardware shall have provisions (holes, lugs, nuts, etc.) so that it can be secured to the support.

Commentary

C2.4.2 Verification and Maintenance

When a new mold is placed into production a complete dimensional check should be made taking into account main dimensions, warping, squareness, flatness, reveals, blockouts and quality of the surface finish. Fixtures and/or templates can aid in checking. The report of this check should be kept on file.

The molds should be reassembled within the dimensional limitations specified for the product on the shop drawings. The overall length, width, thickness, and other basic dimensions should be checked on all sides of the mold. The squareness of the mold should be checked by comparing diagonal measurements to the corners of the mold. Any discrepancy noted in mold dimensional accuracy should be transmitted to production personnel for correction.

A basic assessment of the mold should be made in advance of each casting. Assessment should ensure:

1. The mold has been assembled correctly within specified tolerances.

2. The release agent has been applied satisfactorily and excessive ponding has not occurred in recesses, for example.

3. Concrete splatter, wire ties, dust, and the like have not contaminated the mold surface.

C2.5 Hardware Installation

Inserts should be placed accurately because their capacity depends on the depth of embedment, spacing and distance from free edges. Inserts should also be placed accurately because their capacity decreases sharply if they are not positioned perpendicular to the bearing surface, or if they are not in a straight line with the applied force. It is important to place inserts so that the depth of thread is constant for the same size insert throughout a particular job. Otherwise an erection crew may make mistakes in the field by not always engaging the full thread, (see Fig. C2.5.1). Also, a typical size and thread depth for inserts on projects will minimize the possibility of erection crews using the incorrect size and length of bolts.

Standard

When items cast into units are specified to be plated, galvanized or of stainless steel, a check shall be made to ensure that the material is of the proper type or has the proper coating.

Anchorage shall be provided as indicated on the shop drawings and may be developed through mild steel reinforcing dowels, studs, loops of flat bars, angles, or a combination of these.

When approved by the precast engineer, embedded items such as dowels or inserts, that either protrude from the concrete or remain exposed for inspection, may be installed while the concrete is in a plastic state provided they are not required to be hooked or tied to reinforcement within the concrete and they are maintained in the correct position while the concrete remains plastic. Such items shall be properly anchored to develop the design loads. The concrete surface adjacent to the item shall be flat and well finished for proper bearing and the

Commentary

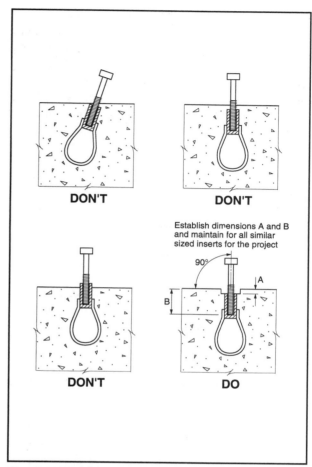

Fig. C2.5.1. Placing of inserts.

The proper anchorage of the insert or hardware is critical. Careful placement of hardware to required tolerances including inclination of protruding bars or structural shapes is important, because the bearing surfaces of the panel hardware and the matching hardware on the structure should be parallel to obtain optimum bearing or load transfer.

Standard

item shall be placed perpendicular to this surface.

Reinforcement shall not be modified, relocated or eliminated to accommodate hardware. If hardware anchors or reinforcing steel cannot be located as shown on the shop drawings, approval for revisions shall be obtained from the precast engineer and all revisions shall be recorded. Connection hardware, carrying load from the member to supports, or handling inserts shall not be plunged or vibrated into concrete already placed in the mold without approval of the precast engineer. If approved by the precast engineer, non-structural hardware items may be installed after the concrete has been placed, and care shall be taken to ensure that concrete around anchorages has been consolidated and that no misplacement of the reinforcing cage has occurred.

Reinforcement extending out of the elements to provide structural connection for sequential casting or cast-in-place concrete shall be located within ±1/2 in. (±12 mm) of shop drawing dimensions unless closer tolerances are indicated. Paste adhering to extended reinforcement shall be removed sufficiently to ensure bonding of bars to concrete that is cast later.

Where angles are used for connections and fastened to the concrete with bolts and inserts, the concrete surface behind the angle shall be flat and well finished for proper bearing. The inserts shall be placed perpendicular to this surface. Such angles may also be anchored with studs or welded to plates in the precast concrete units.

The concrete under plates or angles shall be consolidated in a manner to avoid honeycombing or excessive air voids beneath the plate or angle.

Voids in sleeves, inserts, and anchor slots shall be filled temporarily with readily removable material to prevent entry of concrete during casting. Inserts and sleeves shall be kept clean of dirt or ice by protecting them temporarily with plastic caps or other suitable devices installed in the plant after stripping. Threads on projecting bolts shall be kept free from any deleterious materials and shall be protected from damage and rust.

Stainless steel bolts shall be used to connect stainless steel plates or angles. Carbide inserts shall be used for high speed drilling of holes in stainless steel. Threaded parts of stainless steel bolts shall be lubricated with an antiseize thread lubricant during assembly.

Commentary

Excessive honeycombing beneath the plate or angle may result in diminished structural capacity.

Standard

Dissimilar metals shall not be placed near or in direct contact with each other in moist or saturated concrete unless experience has shown that no detrimental chemical or electro-chemical (galvanic) reactions will occur or surfaces are permanently protected against corrosion. Sleeves, pipes, or conduits of aluminum shall not be embedded in concrete unless they are effectively coated or covered to prevent aluminum-concrete reaction, or electrolytic action between aluminum and steel.

The installation of inserts and fastenings by explosive actuated or power driven tools shall only be allowed where such installation will not damage the structural integrity of the units by being too close to reinforcement or strands, or mar the finish by causing spalling, chipping or cracking of the unit.

Multiple component lifting devices shall be kept matched to avoid non-compatible usage.

When grouped in multiples, lifting loops shall be aligned for equal lifting. Projection of lifting loops for multiple lift point members shall be maintained within a tolerance of ±1/2 in. (±12 mm) from design or plant standard.

Treated or naturally deterioration resistant lumber shall be used, if cast in the concrete. A wood sealer shall be applied to prevent moisture migration from concrete into the wood.

All inserts which are on an exposed face shall be checked to ensure that they are properly recessed to give adequate cover and permit future patching.

2.6 Product Handling

2.6.1 General

All precast concrete units shall be handled in a position consistent with their shape, size, and design to avoid damage or excessive stresses. Units shall be handled and supported only by appropriate devices at designated locations. Lifting devices shall be checked to ensure their locations conform to the shop drawings.

All lifting from threaded inserts shall be with appropriate swivel bolt hardware suitable for the insert unless such lifts are made axial with the inserts, such as by direct bolt fastening to lifting beams or frames. If inclined lifting lines are employed, the angle to the horizontal shall not be less than 45 deg. unless specifically shown by the shop drawings. The bolts being used in the handling inserts shall be of sufficient length to fully engage all

Commentary

Placing wood in concrete should be avoided since the tendency of wood to swell under moist conditions can cause cracking of the concrete even if the wood is sealed.

C2.6 Product Handling

C2.6.1 General

Permanent connection hardware should not be used for handling, except when designed for such additional service and shown on the shop drawings, to prevent damage to the connection or impairment of its performance.

Standard	Commentary

threads plus extending as recommended by manufacturer beyond the threaded position when coil inserts are used. In addition, bolts shall be inspected for cracks, wear or deformations and defective bolts discarded.

Units shall be handled to avoid structural damage, detrimental cracking, architectural impairment or permanent distortion.

Each plant should develop specific unit handling procedures. These should include demolding, lifting, storage, and preparations for transport.

2.6.2 Stripping

C2.6.2 Stripping

Tests shall be performed to confirm that concrete strength meets or exceeds the required values for stripping or detensioning.

All removable inserts, fastenings and mold parts shall be released and/or removed prior to stripping.

Care shall be exercised in removing the precast concrete unit from the mold to prevent damage. The minimum concrete strength, number and location of lifting points for handling of units and details of lifting devices shall be shown on the shop drawings. Units shall only be stripped at points indicated on approved shop drawings or as approved by authorized personnel. Initial lifting shall be made cautiously and gradually and equipment for initial release shall have controls allowing such operation.

Proper rigging procedures should be established and documented. Methods of hooking onto different lifting devices and permissible sling angles should be covered in plant procedures. Where more than two lifting points are used, the lifting should be rigged to allow the planned distribution of pull on lifting devices. Veneered units should be lifted cleanly to prevent chipping or spalling of the veneer edges.

Uniformity of stripping age from unit to unit should be maintained to minimize concrete color hue differences.

A product which is stripped or detensioned prior to achieving adequate concrete strength can crack, spall, or suffer a loss of bond between the strands and the concrete.

Significant member stresses are often produced during demolding. Proper care should be taken to minimize demolding stresses and keep them within acceptable limits.

Individual precast concrete elements shall be clearly marked with identification as shown on the shop drawing, including date of casting or serially numbered.

Panel markings may be made in the wet concrete, painted on the units, or they may be made by tagging as long as the tags are securely fastened to the units. In all cases the markings shall remain legible for the longest storage period that can be anticipated.

Surface of unit shall be checked for any dents or marks that may be the result of mold wear or deterioration. Also, unit shall be checked for voids or excessive air holes that would indicate improper vibration.

A system for evaluating product deficiencies should be established so that problems and their sources can be identified. Problems which occur when products are stripped cannot be resolved easily if they are not identified prior to product storage.

2.6.3 Yard Storage

C2.6.3 Yard Storage

Storage shall be planned carefully to ensure deliv-

Units are generally stored with two-point supports spaced ap-

Standard

ery and erection of the units in an acceptable condition. Units shall be supported and stored on unyielding supports at designated blocking point locations. Units shall be stored on a firm, level and well-drained surface with identification marks visible. Dunnage and storage racks, such as A-frames and vertical racks, shall be well constructed and aligned to ensure that the precast concrete units are supported in a given plane to minimize warping, bowing or cracking of the units and stabilized against potential lateral loads.

Dunnage shall be placed between products themselves as well as between products and storage devices. Such dunnage shall be of a material and type that will not create stacking or staining marks or otherwise cause damage to the finished products. Where necessary, storage apparatus shall be protectively covered to avoid any discoloration or staining of the finished products.

Embedded items and sleeves shall be protected from penetration of water or snow during cold weather.

2.6.4 Cleaning

All surfaces of the precast concrete units to be exposed to view shall be cleaned, as necessary, prior to shipping to remove dirt and stains which may be on the units. The cleaning procedures shall not detrimentally affect the concrete surface finish.

2.6.5 Loading

Units shall be loaded as necessary:

1. to permit their removal for erection from the load in proper sequence and orientation to minimize handling.

2. with proper supports, blocking, cushioning, and tie-downs to prevent or minimize in-transit damage.

    All blocking, packing and protective materials shall be of a type that will not cause damage,

Commentary

proximately at the fifth point. Two-point support is recommended because if support is continuous across three or more points, the member may bridge over one of the supports (because of differential support movement) and result in bowing and cracking.

Proper member support during storage will minimize warping and bowing. Warpage in storage may be caused by temperature or shrinkage differential between surfaces, creep and storage conditions. Warpage and bowing cannot be totally eliminated, although it can be minimized by providing blocking so that the unit remains plane.

Units stored leaning on one another may induce high stress loads in long storage lanes. The "domino effect" (cumulative loading) should be considered. Units should be stacked against both sides of the supports to equalize loading and to avoid overturning.

Protective material should be provided at points of bearing and contact with exposed surfaces. Care should be taken to prevent surface staining and chipping or spalling of the edges and corners of the units. All blocking, packing, and protective materials should be of a type that will not cause damage, staining, or objectionable disfigurement of the units. Staggered or irregular blocking should be avoided. When setting one unit against another, non-staining, protective blocks should be placed immediately in line with the supports of the first unit.

The precast concrete units should be protected from contact with earth, oil, gas, tar, or smoke.

C2.6.4 Cleaning

A small area should be cleaned and appraised to be certain there is no adverse effect on the surface finish before proceeding with the work.

For information on removing specific stains from concrete, reference should be made to "Removing Stains and Cleaning Concrete Surfaces," IS 214, published by the Portland Cement Association, Skokie, IL.

C2.6.5 Loading

Protective covering of the units during transportation is normally not required.

Standard

staining or other disfigurement of the units.

The blocking points and orientation of the units on the shipping equipment shall be as designated on the shop drawings.

3. with proper padding between units and chains or straps to preclude chipping of edges or damage to returns.

2.7 Surface Finishes

2.7.1 General

Each plant shall develop quality requirements for all architectural finishes prior to undertaking actual production of such finishes. Such requirements shall include samples and production procedures. A finishing process shall produce an acceptable uniform appearance without detriment to required material properties.

All finishes of precast concrete units shall be stated on shop drawings. Reference samples or mockup units shall be available in the plant so that all concerned can be assured that standards of finish and exposure are being maintained.

Commentary

C2.7 Surface Finishes

C2.7.1 General

Concrete is a variable material and, even after final finishing, there will be a range in color or texture. Some differences are to be expected. Finishing techniques vary considerably between individual plants. Many plants have developed specific techniques supported by skilled operators or special facilities. In addition, many finishes cannot be achieved with equal visual quality on all faces of the unit. The reason for this comprises several factors such as mix proportions, variable depths (and pressures) of concrete, and small differences in consolidation techniques, particularly in the case of intricate shapes with complex flow of concrete.

It should be recognized that some blemishes or variations in color occur in architectural precast concrete. For example, units containing aggregates and matrices of contrasting colors will appear less uniform than those containing materials of similar colors. Consistency in apparent color of all finishes can be enhanced by color compatibility of materials. If the coarse aggregate, fine aggregate and cement paste are similar in color, the depth of exposure and "patchy" effects (minor segregation of aggregate) will not be as noticeable in maintaining color consistency. In contrast, if material colors are quite different, products may appear blotchy for the same reason.

Further, form-finished units may appear less uniform in color than the same units subsequently given an exposed aggregate finish. Uniformity in color, even within small units finished with black coloring, and for units using gray or buff cement, may vary from unit to unit. The color of precast concrete can vary between adjacent elements due to daily variations in the water-cement ratio and curing conditions for the concrete. This is less pronounced in mixes using white cement than those using gray or buff cement. Additionally, this color variation will be less pronounced in finishes that have some degree of aggregate exposure. The degree of uniformity normally improves with an increased depth of exposure. Buff color achieved with fine aggregate is more easily controlled even in combination with gray cement. When fabrication continues over extended periods, color can vary because of the changes in the physical characteristics of cements, coarse aggregates and sands, even though they may be from the same sources.

The effect of gravity during consolidation forces the larger aggregates to the bottom and the smaller aggregates, plus the sand and cement content, upwards. Consequently, the down-face in the mold will nearly always be the most uniform and dense surface of the unit. The final orientation of aggregates

Standard

The extent to which aggregates are exposed or "revealed" shall be no greater than one-third the average diameter of the coarse aggregate particles, or not more than one-half the diameter of the smallest sized coarse aggregate.

A demarcation feature shall be incorporated into the surface of a unit having two or more different mixes or finishes. The different face mixes shall have reasonably similar behavior with respect to shrinkage in order to avoid cracking at the demarcation feature due to differential shrinkage.

Appearance, color, and texture of surface finish of all units shall match within the acceptable range of the color, texture, and general appearance of the approved sample panels.

2.7.2 Smooth

The molds shall be carefully made and finished so as to present a smooth, unmarked surface. If air voids are anticipated on return surfaces, the sample shall be used to establish acceptability of such voids with respect to frequency, size and uniformity of distribution of the voids.

Commentary

may also result in differences in exposure between the downface and returns in exposed aggregate surfaces. Emphasis should be placed on choosing suitable concrete mixes with aggregates which are reasonably spherical or cubical to minimize differences. For large returns, or situations where it is necessary to minimize variations in appearance, concrete mixes should be selected where the aggregate gradation can be uniformly controlled and preferably fully graded. Exposures should be medium to deep and color differences between the ingredients of the mix should be minimal.

Panels with large returns may also be cast in separate pieces and joined with dry joints. This enables all faces to be cast with the same orientation. If this is the indicated production method, attention should be paid to suitable fillets and reinforcement at the corners, and a groove or architectural feature should be incorporated at the dry joint. Although the dry joint may not show with certain mixes and textures, a groove is generally required to help mask it.

As a general rule, a textured surface is aesthetically more satisfactory than a smooth surface because the texture of the surface to a very large extent camouflages subtle differences in texture and color of the concrete.

C2.7.2 Smooth

A smooth off-the-form finish may be one of the most economical, but is perhaps the most difficult to produce, as the color uniformity of gray, buff, or pigmented surfaces is extremely hard to achieve. The cement exerts the primary color influence on a smooth finish because it coats the exposed surface. In some instances the sand may also have some effect. Initially, this is unlikely to be significant unless the sand contains a high percentage of fines or is itself highly colored. As the surface weathers, the sand will become more exposed, causing its color to become more pronounced. The color of the coarse aggregate should not be significant unless the particular unit requires extremely heavy consolidation. Under this circumstance, some aggregate transparency may occur, causing a blotchy, non-uniform appearance.

Aggregate transparency or "shadowing" is a condition in which a light colored, formed concrete surface is marked by dark areas similar in size and shape to particles of dark or deeply colored coarse aggregate in the concrete mix. When encountered, it usually appears on smooth surfaces.

Standard

2.7.3 Sand or Abrasive Blast

Uniformly texturing a panel by sand or abrasive blasting requires trained operators. The type and grading of abrasives utilized during the blasting process shall remain the same throughout the entire project.

Commentary

Formwork for smooth-surfaced concrete is perhaps the most critical and the most difficult to control of any type of formwork encountered for precast concrete, particularly where large single-plane surface areas are involved. Any imperfection in the surface of the mold or any misalignment is immediately apparent and becomes the predominant factor in the character of the surface. An impervious surface such as plastic liners, steel, overlaid plywood, or fiberglass surfaced plywood will usually result in a lighter color and more uniform appearance if joints have been properly prepared. In general, the joints of the materials used to construct the casting surfaces are difficult to hide.

The smooth cement film on the concrete may be susceptible to surface crazing, i.e., fine and random hairline cracks, when exposed to wetting and drying cycles. This is, in most cases, a surface phenomenon and will not affect structural properties or durability. In some environments, crazing will be accentuated by dirt collecting in these minute cracks. This will be more apparent in white than gray finishes and in horizontal more than vertical surfaces.

When air voids of a reasonable size, 1/8 to 1/4 in. (3 to 6 mm) are encountered on return surfaces, it may be desirable to retain them rather than filling and sack rubbing them in. Color variations can occur when sacking is performed.

Even with good quality control, smooth finished concrete will exhibit some negative aesthetic features such as color variations, air voids, minor surface crazing, and blotchiness, especially on non-profiled flat panels. Repairs to this finish tend to be even more noticeable after weathering.

C2.7.3 Sand or Abrasive Blast

Sand or abrasive blasting of surfaces is suitable for exposure of either large or small aggregates. Uniformity of depth of exposure between panels and within panels is essential in abrasive blasting, as in all other exposed aggregate processes, and is a function of the skill and experience of the operator. As much as possible, the sandblasting crew and equipment used should remain the same throughout the job. The type and grading of abrasives affect the resulting surface finish. Different shadings and to some extent, color tone will vary with depth of exposure. The age of the panel at time of blasting will also affect the speed and depth of exposure. The age and strength of the concrete at time of blasting should be consistent throughout the project in order to achieve the desired uniform finish and color from panel to panel.

The degree of uniformity obtainable in a sandblasted finish is generally in direct proportion to the depth of sandblasting. A light sandblasting may look acceptable on a small sample, but uniformity is rather difficult to achieve in a full size unit. A light sandblast will emphasize visible defects, particularly bugholes, and reveal defects previously hidden by the surface skin of the concrete. The lighter the sandblasting, the more critical the skill of the operator, particularly if the units are sculptured. Small variances in concrete strength at time of blasting may further complicate results.

Sculptured units will have air voids on the returns which might show strongly in a light sandblasted texture. If such air holes are of a reasonable size, 1/8 to 1/4 in. (3 to 6 mm), it is strongly rec-

Standard

Sands used for blasting operations shall be free of deleterious substances such as fine clay particles. If sand is used as the abrasive, a high silica content sand shall be used rather than bank or river sand. The abrasive used shall not cause any color contamination.

2.7.4 Acid Etched

Acid etching may be accomplished by: (1) brushing the surface with a stiff bristled fiber brush immersed in the acid solution; (2) spraying acid and hot water onto the panel surface using specially designed pumps, tanks and nozzles; or (3) immersing the unit for a maximum of 15 minutes in a tank containing from 5 to 35 percent hydrochloric acid. In sandwich wall panels, where the insulation is exposed at the edges of the panel only those methods that prevent damage to the insulation shall be used.

An acceptable range of concrete temperatures and strength levels during the application of the acid shall be established to provide uniform finish quality.

In cases where aggregates are to be exposed to considerable depth, only acid resistive siliceous aggregates shall be used. The concrete unit shall be well wetted with clean water prior to acid treatment because acids will penetrate faster and deeper into dry concrete. Acid shall not be allowed to lie on the surface longer than 15 min. Deep etch shall be achieved by multiple treatments. After completion of acid etching, the unit shall be thoroughly flushed with water.

Commentary

ommended that they be accepted as part of the texture because filling and sack-rubbing may cause color variations.

The objective of a brush-blast, which is little more than a uniform scour cleaning that lightly textures the surface skin, is to remove minor surface variations. A brush-blast surface seldom appears uniform at close inspection and should be viewed at a distance for uniformity. Brush-blast is only used: (1) on reveals and other small areas for accents due to its non-uniform appearance on large areas or (2) to improve bond of coatings to unit..

C2.7.4 Acid Etched

Acid etching is most commonly used for light or medium exposure dissolving the surface cement paste to reveal the sand with only a small percentage of coarse aggregate being visible. Acid etching of concrete surfaces will result in a fine, sandy texture if the concrete mix and its consolidation have produced a uniform distribution of aggregates and cement paste at the exposed surfaces. Concentrations of cement paste and under and over etching of different parts of a concrete surface, or variation in sand color or content may cause some uniformity problems, particularly when the acid etching is light or used for large, plain surfaces. Carbonate aggregates e.g., limestones, dolomites and marbles, may discolor or dissolve due to their high calcium content.

With lighter textures, color compatibility of the cement and the aggregates becomes more important to avoid a blotchy effect. White or light colors are forgiving to the eye and increase the likelihood of better color match from unit to unit.

There is a minimum depth of etch that is required to obtain a uniform surface. To attempt to go any lighter than this will result in a blotchy panel finish. This depth will expose sand and only the very tip of the coarse aggregate. It is difficult to achieve a totally uniform very light exposure on a panel that is highly sculptured. This is due to the acid spray being deflected to other areas

Standard

Prior to acid etching, all exposed metal surfaces, particularly galvanized metal, shall be protected with acid-resistant coatings. These include vinyl chlorides, chlorinated rubber, styrene butadiene rubber (not latex), bituminous paints and enamels and polyester coatings.

Touch-up of all exposed galvanized metals cast in the precast concrete units affected by acid washing or etching shall be done utilizing a 3-mil thickness of a single component zinc-rich compound with 95 percent pure zinc in the dried film.

2.7.5 Retarded

Surface retarders that are to be used to expose the aggregate shall be thoroughly evaluated prior to use. A sample panel shall be made to determine the effects created by the mold and concrete materials. This involves using the particular type of cement, aggregate, and specific mix selected for the product.

When using a retarder, the manufacturer's recommendations shall be followed. Surface retarders shall be applied by roller, brush, or spray, and care shall be taken to ensure uniform application of retarders to the mold surface. Water shall not contact the retarder on the mold surface before the concrete is placed to prevent activation of the retarder.

Commentary

of the panel, particularly at inside corners. This may be acceptable if the sculpturing creates differential shadowing.

Prewetting the concrete with water fills the pores and capillaries and prevents the acid from etching too deeply, and also allows all acid to be flushed after etching. Older dried concretes are likely to be more carbonated. Although the reactions of carbonates with the acid might not be much faster than those with other cement compounds, they cause greater effervescence so that the reaction is far more obvious and seems to be going faster.

Acid solutions lose their strength quickly once they are in contact with cement paste or mortar; however, even weak, residual solutions can be harmful to concrete due to possible penetration of chlorides. Failure to completely rinse the acid solution off the surface may result in efflorescence or other damaging effects.

C2.7.5 Retarded

Retardation involves the application of a specialized chemical to the concrete surface (normally the mold surface) that delays the surface cement paste from hardening within a time period and to a depth depending upon the type or concentration of retarder used.

Chemical retarders are available for the face-down or face-up methods of casting, and for horizontal as well as vertical surfaces. Retarders are available for light, medium and deep exposures. The degree of uniformity normally improves with an increased depth of exposure.

The effectiveness of the retarder will vary as it is extremely sensitive to changes in the rate of hydration due to different temperatures, humidity or water content of the face mix. The depth of reveal or retardation will be deeper: (1) the wetter the mix; (2) the slower the time of set; (3) the more aggregate in the mix; and (4) the closer together the coarse aggregate.

Retarders function by delaying, not preventing, the set of a given amount of cement paste in order that the aggregate can be easily exposed. This concept will help in analyzing various mix designs for depth of retardation. If more sand or coarse aggregate is added to a mix; with proper consolidation, there will be less cement paste per volume of material at the surface, thus a deeper exposure.

Standard

The retarded surface shall be exposed by removing the matrix material to match the approved sample.

2.7.6 Tooled or Bushhammered

Operators shall be trained to produce a uniformly textured surface to match the approved sample when exposing aggregate by tooling or bushhammering

2.7.7 Honed or Polished

Care shall be taken to obtain a uniform depth of grind from unit to unit to minimize finish variations. An approved sample unit shall be kept near the grinding for comparison and evaluation of the product finish. Air voids in the concrete surface shall be filled before each of the first few grinding operations or no later than half-way through the third grinding step using a sand-cement mixture that matches the matrix of the concrete. Careful filling and curing are required and the next grinding operation shall not be performed until the fill material has reached sufficient strength.

Commentary

Some retarders are effective for long periods of time while others are active for only a few hours. Water in contact with the retarder before the concrete is placed activates the retarder's action prematurely and may result in a non-uniform surface.

The retarded concrete should be removed the same day that the units are stripped. Any delay in removing the matrix will result in a lighter, less uniform etch. Preliminary tests should be performed before planning the casting for a large project to determine the most suitable finishing time. The timing of the surface finishing operation should be consistent each day as some retarders cease to delay the hardening process as the product cures.

C2.7.6 Tooled or Bushhammered

Concrete may be mechanically spalled or chipped with a variety of hand and power tools to produce an exposed aggregate texture. The technique usually is called tooling or bushhammering and is most suitable for flat or convex surfaces. Pneumatic or electric tools may be fitted with a bushhammer, comb chisel, crandall, or multiple pointed attachments. The type of tool will be determined by the surface effect desired. Hand tools may be used for small areas, corners, and for restricted locations where a power tool cannot reach.

Orientation of equipment and direction of movement for tooling should be kept uniform throughout the tooling process as tooling produces a definite pattern on the surface. Variations due to more than one person working on the panels may occur with this finish. Care should be exercised to avoid exerting excessive pressure on the tool, especially when starting, so as not to remove more material than either necessary or desirable.

Bushhammering at outside corners may cause jagged edges. If sharp corners are desired, bushhammering should be held back from the corner. It is quite feasible to execute tooling along specific lines. If areas near corners are to be tooled, this should normally be done by hand since tools will not reach into inside corners. Chamfered corners are preferred with tooled surfaces and with care a 1 in. (25 mm) chamfer may be tooled.

C2.7.7 Honed or Polished

The grinding of concrete surfaces produces smooth, exposed aggregate surfaces. Grinding is also called honing and polishing depending on the degree of smoothness of the finish. In general, honed finishes are produced by using grinding tools varying from about No. 24 coarse grit to a fine grit of about No. 220 to produce a smooth but matte finish free of pits and scratches. Polishing is accomplished after honing. Polishing consists of several successive grinding steps, each employing a finer grit than the preceding step, then a buffer brick or felt pad with tin oxide polishing powder is used to produce a high gloss polish.

Mold flatness should be ±3/32 over 13 ft (±2 mm over 4 m). For a very large project requiring many castings it could be viable to have the mold surface machined flat during mold manufacture since minor variations in flatness can increase cutting time during initial polishing operations. In addition, panels should be placed onto pre-levelled blocks with the final alignment carried out using a laser beam, then wedged for final adjustment.

Standard

Care shall be exercised in the application of polishing compounds to prevent creating a visually unacceptable halo effect on the surface, particularly if applied manually over a portion of surface during the blending of defective spots.

Compressive strength of the concrete shall be 5000 psi (35 MPa) before starting any honing or polishing operations. The concrete shall have a uniform and dense surface. When choosing aggregates special consideration shall be given to their hardness.

2.7.8 Form Liner

Form liner panels shall be secured in molds by methods which will not permit impressions of nail heads, screw heads, rivets or the like to be imparted to the surface of the concrete unless this feature is desired. Attempts shall be made to camouflage anomalies to within the pattern of the texture.

An investigation shall be made to determine whether staining or discoloration may occur from the liner material, its fastenings, or joint sealers. Care shall be taken to use mold release agents and retarders that are compatible with the liner and the liner shall not be subjected to temperatures in excess of those recommended by the manufacturer.

2.7.9 Veneer Facing Materials

Quality requirements (design and production procedures) for finishes derived from materials such as natural cut stone (granite, limestone, marble), brick, ceramic or quarry tile, and architectural terra cotta shall be based on previous records with the identical materials, or sufficient testing of sample and

Commentary

Uniformity of appearance should not be a problem on flat cast faces but vertical faces are likely to show variation in aggregate density. Careful consideration should be given to manufacturing methods when panel returns do not align with adjacent window glass but abut with flat panels as this will highlight the textural differences.

When a 90 deg. return of a panel is honed or polished, it may prove beneficial to sequentially cast the return to allow it to be cast in a horizontal position which will create a more dense uniform surface.

C2.7.8 Form Liner

Form liners may be incorporated in or attached to the surface of a mold to produce the desired pattern, shape or texture in the surface of the finished units. The method of attaching the form liner should be studied for resulting visual effect.

A form liner texture can be of considerable influence in assisting as-cast surfaces to appear more uniform. Form liner material selection depends on the amount of usage and whether or not the pattern has undercut (negative) drafts. Matching joints between liners, is very difficult. Liners should either be limited to widths less than the available width of the liner, or liner joints should be at form edges or be detailed as an architectural feature in the form of a groove, recess or rib.

C2.7.9 Veneer Facing Materials

Color control or blending for uniformity should be done in the stone fabricator's plant since ranges of color and shade, finishes, and markings such as veining, seams and intrusions are easily seen during the finishing stages. A qualified representative of the owner who understands the aesthetic appearance requested by the owner or architect should perform this color control. Acceptable color of the stone should be judged for an

Standard

mockup units to establish performance criteria under the service conditions. Particular attention shall be paid to the compatibility of materials with respect to differential expansion and contraction caused by thermal and moisture changes; and also the differential volume change (shrinkage) between the veneer facings and concrete backup. If the materials do not have similar physical properties, the final design shall include compensation for some interaction of the different materials.

Natural stone. Cut stones that are easily stained by oils and rust, require the molds to be lined with polyethylene sheets or other non-staining materials.

A complete bondbreaker between natural stone veneer and concrete shall be used. Bondbreakers shall be one of the following: (1) a liquid bond breaker applied to the veneer back surface prior to placing the concrete; (2) a minimum 4 mil polyethylene sheet; or (3) a 1/8 in. (3 mm) polyethylene foam pad or sheet. The bondbreaker shall prevent concrete from entering the spaces between pieces of veneer and thereby potentially inhibiting differential movements. Connection of the veneer to the concrete shall be with mechanical anchors which can accommodate some relative movement. Preformed anchors, fabricated from Type 302 or 304 stainless steel shall be used. Close supervision is required during the insertion and setting of the anchors. If the anchor is placed in epoxy, it shall not be disturbed while the epoxy sets.

When using epoxy in anchor holes, 1/2 in. (12 mm) long compressible rubber or elastomeric grommets or sleeves shall be used on the anchor at the back surface of the stone, and the epoxy manufacturer's recommendations as to mixing and curing temperature limitations shall be followed.

The strength of the stone veneer material shall be known or determined along with that of the anchorage system to assure adequate strength to resist stresses during handling, transportation, erection, and service conditions.

Veneer joints within a concrete element shall allow for differential movement between materials. In the mold, the veneer pieces shall be temporarily spaced with a non-staining, compressible spacing material, which will not adversely affect the sealant to be applied later. Shore A hardness of the gasket shall be less than 20 durometer.

Commentary

entire building elevation rather than for individual panels.

All testing to determine the physical properties of the stone veneer with the same thickness and finish as will be used on the structure should be conducted by the owner prior to the award of the contract. This will reduce the need for potentially costly repairs or replacement should deficiencies in the stone veneer be found after start of fabrication.

The physical properties of the clay products should be compared with the properties of the concrete backup. These properties include the coefficient of thermal expansion, modulus of elasticity and volume change due to moisture.

Standard

The gaskets shall be of a size and configuration that will provide a pocket to receive the sealant and also prevent any of the concrete backup from entering the joints between the veneer units. Spacer materials shall be removed after the panel has been stripped from the mold unless a resilient sealant backup is utilized.

When stone veneer is used as an accent or feature strip on precast concrete panels, a space shall be left between the edge of the stone and the precast concrete to allow for differential movements of the materials. This space shall be caulked as if it were a conventional joint.

Clay products. Bricks with an initial rate of absorption (suction) of less than 30 g per 30 sq in. per min (30 g per 194 cm^2 per min), when tested in accordance with ASTM C67, are not required to be wetted. However, brick with high suction or with an initial rate of absorption in excess of 30 g per 30 sq in. per min (30 g per 194 cm^2 per min) shall be wetted prior to placement of the concrete. Unglazed quarry tile and frost-resistant glazed wall tiles, generally, are not required to be wetted. Terra cotta units shall be soaked in water for at least one hour prior to placement to reduce suction and be damp at the time of concrete placement.

Because variations in brick or tile color will occur, the clay product supplier shall preblend any color variations and provide units which fall within the color range selected by the architect. Clay products that suffer from various surface defects, such as chips, spalls, face score lines and cracks shall be culled from the bulk of acceptable units by the clay product supplier or precaster according to the architect's requirements and applicable ASTM specifications.

Ceramic glaze units, where required for exterior use, may craze from freeze-thaw cycles or the bond may fail on exposure; therefore, the manufacturer shall be consulted for suitable materials and test data backup.

Clay product faced units shall have joint widths controlled by locating the units in a suitable template or grid system set out accurately on the mold face. If an elastomeric form liner is used, it shall be produced to a tolerance of ±1/32 in. (±0.8 mm).

In addition to normal dimensional checks, clay product coursings shall be rechecked for alignment and all clay products shall be fully seated against the mold face. Clay product units shall be checked for tight fit and wedged if not tightly secured - espe-

Commentary

Clay products. Whole bricks are generally not used in precasting because of the difficulty in adequately grouting the thin joints and the resultant necessity to use mechanical anchors.

Clay products with high suction or with a high initial rate of absorption should be wetted prior to placement of the concrete to reduce the amount of mix water absorbed and thus improve bond.

Some bricks are too dimensionally inaccurate for precast concrete applications. They may conform to an ASTM specification suitable for site laid-up applications, but they are not manufactured accurately enough to permit their use in a preformed grid that is used to position bricks for a precast concrete

Standard

cially on return sections - to prevent grout leakage to the exposed face of the panel. Concentrations of clay products of the same shade shall be avoided, and chipped or warped units shall be checked and removed.

Care shall be taken during mortar or concrete placing and consolidation to prevent movement of the individual facing materials, which would upset the appearance of the finished surface.

After the concrete cures and the unit is removed from the mold, joints shall be filled, if necessary, with pointing mortar or grout carefully formulated for color and texture. Before pointing, joints shall be saturated with clean water. After the joints are properly pointed and have become thumbprint hard, they shall be: (1) tooled to a smooth concave surface, which offers the best durability, or (2) struck and troweled flush with the face of the clay units. Initial grout cleanup shall be done within 15 minutes of pointing to avoid hard setting of the grout on the units. Final cleanup shall be completed within 60 to 90 minutes.

Thin brick and ceramic tile shall be applied to a recessed concrete surface that has been properly roughened by sandblasting or bushhammering using dry-set mortar conforming with ANSI A118.1 or latex-portland cement mortar conforming with ANSI A118.4. Installation using either dry-set or latex-portland cement mortars shall conform to ANSI A108.5. When dry-set mortar is used, the necessity of wetting either the concrete surface or clay product is eliminated. Since lattices vary considerably, each latex manufacturer's directions shall be followed explicitly, particularly with regard to curing.

Units shall be grouted and tooled using dry-set or latex-portland cement grouts conforming with material and installation specifications contained in ANSI A118.6 and ANSI A108.10.

2.7.10 Sand Embedded Materials

Bold and massive, rock-like architectural qualities may be achieved by hand placing large diameter stones (cobbles or boulders), fieldstone or flagstone into a sand bed or other special bedding material. The depth of the bedding material shall keep the backup concrete 25 to 35 percent of the stone's diameter away from the face.

Extreme care shall be taken to ensure that the aggregate is distributed evenly and densely on all sur-

Commentary

unit. Tolerances in an individual brick of ±3/32 in. (±2 mm) or more cause problems. Brick may be available from some suppliers to the close tolerances [±1/16 in. (±2 mm)] necessary for precasting. Close tolerances also can be obtained by saw cutting each brick, but this substantially increases cost.

C2.7.10 Sand Embedded Materials

The sand embedment technique reveals the facing material and produces the appearance of a mortar joint on the finished panel.

Standard

faces, particularly around corners, edges and openings. To achieve uniform distribution and exposure all aggregate shall be of one size gradation. Where facing materials are of mixed colors, their placement in the mold shall be carefully checked for the formation of unintended patterns or local high incidence of a particular color. If it is the intention to expose a particular facet of the stone, placing shall be carefully checked with this in mind before the backup concrete is placed.

2.7.11 Unformed Surface Finishes

For unformed surfaces, visible (exposed) areas shall have finishes which are realistic in relation to the exposure, production techniques, unit configuration, and quality requirements. The finish requirements for all unformed surfaces shall be shown on the shop drawings.

Troweling shall not be done on a surface that has not been floated. Magnesium screeds and floats shall be used with air-entrained or lightweight concrete to minimize surface rippling, tearing, or pullouts.

To obtain a durable surface on unformed concrete, proper procedures shall be carefully followed. Surfaces shall be filled and struck off immediately after concrete placing and consolidation and then rough floated. The operations of screeding, floating and first troweling shall be performed in such a manner that the concrete will be worked and manipulated as little as possible in order to obtain the desired result. Over-manipulation of the surface shall be avoided. Overworking of the surface of structural lightweight aggregate concrete shall be avoided even more so than with normal weight concrete.

Each step in the finishing operation from bull floating to the final floating or troweling, shall be delayed as long as possible to permit the desired grade and surface smoothness to be obtained. If excess moisture or bleed water does accumulate, it shall either be allowed to evaporate or, if excessive, it shall be removed by blotting with mats, draining, or pulling off with a loop of hose, so the surface can lose its water sheen before the next finishing operation is performed. Under no circumstances shall any finishing tool be used in an area before accumulated water has been removed, nor shall neat cement or mixtures of sand and cement be worked into the surface to dry such areas. During final troweling, the surface shall be manipulated only as required to produce the specified finish and to close any surface cracks that may

Commentary

C2.7.11 Unformed Surface Finishes

The back of a precast concrete unit may be given a variety of finishes depending on the intended service or appearance. These may include a screed, light broom, float, trowel, stippled, or water-washed or retarded exposed aggregate finish.

Over-manipulation of the concrete surface brings excessive fines and water to the top which impairs the quality of the finished surface causing such undesirable effects as checking, crazing, dusting, and discoloring.

Excess moisture or bleed water is not as likely to appear and accumulate between finishing operations if proper mix proportions and consistency are used.

Standard

have developed.

Prior to initial set of the concrete, a check shall be made to ensure that the floating or finishing operation do not result in high areas or ridges around plates or inserts that have been cast into the unit. Also, screeding of units shall be checked to ensure uniform thickness across the entire unit.

2.7.12 Applied Coatings

Whenever concrete is to be painted or stained, only mold release agents compatible with coating shall be permitted unless surface preparation is required to assure good adhesion between the coating and the concrete.

Coating applied to exterior surfaces shall be of the breathing type, (permeable to water vapor but impermeable to liquid water). The coating manufacturer's instructions regarding mixing, thinning, tinting, and application shall be strictly followed.

Whenever concrete is so smooth that it makes adhesion of some coatings difficult to obtain, such surfaces shall be lightly sandblasted, acid etched, or ground with silicon carbon stones to provide a slightly roughened, more bondable surface.

2.8 Repairs

Plant personnel involved with repairs shall have detailed written repair procedures.

When objectionable discrepancies do occur, immediate action shall be taken to determine the cause of such discrepancies and apply solutions to prevent their further occurrence.

Since the techniques and materials for repairing architectural precast concrete are affected by a variety of factors including mix ingredients, final finish, size and location of damaged area, temperature and humidity conditions, age of member, surface texture, etc., precise methods of repairing cannot be detailed in this Manual.

Commentary

C2.7.12 Applied Coatings

Paints may be used for purely decorative reasons. See Article 2.10 for discussion on sealers, both clear and pigmented. Every paint is formulated to give certain performance under specific conditions. Since there is a vast difference in paint or stain types, brands, prices, and performances, knowledge of composition and performance standards is necessary for obtaining a satisfactory concrete paint or stain.

C2.8 Repairs

The written repair procedure should clearly define when quality control is required to consult with the precast engineer to determine an appropriate fix or repair. Major repairs should not be attempted until an engineering evaluation is made to determine whether the unit will be structurally sound, and, if so determined, the repair procedure should be approved by the precast engineer.

A certain amount of product repair is to be expected. Repair work requires expert craftsmanship and careful selection and mixing of materials, if the end result is to be structurally sound, durable and aesthetically pleasing. Repairs are acceptable provided the structural adequacy, serviceability, and the appearance of the product are not impaired. Excessive variation in color and texture of repairs from the surrounding surfaces may result in the panels not being approved until the variation is minimized.

Repairs should be done which ensure that the repaired area will conform to the balance of the work with respect to appearance, structural adequacy and durability. Repairs should be made at the plant well in advance of shipping to allow for proper curing of repaired area. Provision should be made to repair damaged products in either the finish or curing area of the plant or a special area set aside for this purpose. Repairs should be avoided in the yard storage area or on the truck just prior to shipment. Slight color variations can be expected between the repaired

Standard

Adequate curing methods for repairs shall be implemented as soon as possible to ensure that the repair does not dry out too quickly. Corrosion protected materials shall be touched-up upon completion of all intended curing and acid cleaning.

All repaired products shall be inspected by quality control personnel to ensure that proper repair procedures including curing have been followed and that the results are acceptable. Repairs shall be evaluated after having been cured.

2.9 Acceptability of Appearance

Uniformity of color and intensity of shading are generally a matter of subjective individual judgment. Therefore, it is beyond the scope of this Standard to establish definitive rules for product acceptability on the basis of appearance. The finished face surface shall have no obvious imperfections other than minimal color and texture variations from the approved samples or evidence of repairs when viewed in good typical daylight illumination with the unaided naked eye at a 20 ft (6 m) viewing distance. Appearance of the surface shall not be evaluated when light is illuminating the surface from an extreme angle as this tends to accentuate minor surface irregularities.

Unless approved otherwise in the sample/mockup process (see Article 1.5.4) the following is a list of finish defects that shall be properly repaired, if obvious when viewed at a 20 ft (6 m) distance.

1. Ragged or irregular edges.

Commentary

area and the original surface due to the different age and curing conditions of the repair. Time (several weeks) will tend to blend the repair into the rest of the member so that it should become less noticeable. Gross variation in color and texture of repairs from the surrounding surfaces will require removal of the repair material and reapplication of new repair material. Small cracks, under 0.010 in. (0.25 mm), may not need repair, unless failure to do so can cause corrosion of reinforcement. If crack repair is required for the restoration of structural integrity or member finish, cracks may be filled or pressure injected with a low viscosity epoxy.

To match specified architectural finishes, repair mixes should be developed early, following approval of initial sample. A trial and error process is normally required for each newly developed face mix to effectively match color and texture.

C2.9 Acceptability of Appearance

It should be stated in the contract documents who the accepting authority will be — contractor, architect, engineer of record, owner or jobsite inspector.

At the time the visual mockups or initial production units are approved, the acceptable range in color, texture and uniformity should be determined.

1. It is strongly recommended that all edges of precast concrete units be detailed with a reasonable radius or chamfer, rather than leaving them as sharp corners. Sharp corners chip easily during handling and during service in the building. It can be difficult to cast concrete to a 45 degree point because of the size of the aggregates. When the edge is sharp, only fine aggregate collects there and this weakens the edge. Also, voids occur due to the interference of larger aggregate. Therefore, this edge should have a cutoff or quirk. The size of the quirk return should not be less than 3/4 in. (20 mm), nor less than 1.5 times the maximum aggregate size used in the concrete mix.

Standard

2. Excessive air voids (commonly called bug-holes) larger than 1/4 in. (6 mm) evident on the exposed surfaces.

3. Adjacent flat and return surfaces with greater texture and/or color differences than the approved samples or mockups.

4. Casting and/or aggregate segregation lines evident from different concrete placement lifts and consolidation.

5. Visible mold joints or irregular surfaces.

6. Rust stains on exposed surfaces.

7. Units with excessive variation of texture and/or color from the approved samples, within the unit or compared with adjacent units.

Commentary

2. Sculptured panels, channel panels, and panels with deep returns may have visible air voids on the returns. These air voids or "bug/blow holes," become accentuated when the surface is smooth, acid-etched or lightly sandblasted. If the air holes are of a reasonable size, 1/8 to 1/4 in. (3 to 6 mm), it is recommended that they be accepted as part of the texture. Filling and sack-rubbing could be used to eliminate the voids. However, this procedure may cause color differences. Samples or the mockup panel should be used to establish acceptable air void frequency, size, and distribution.

3. Returns in some finishes will not appear exactly like the front face (down-face) due to a number of factors such as mix proportions, variable depths (and pressures) of concrete, and small differences in consolidation techniques, particularly in the case of intricate shapes with complex flow of concrete. The effect of gravity during consolidation forces the large aggregates to the bottom and the smaller aggregates, plus the sand and cement content, upwards. Consequently, the down-face in the mold will nearly always be more uniform and denser than the returns or upper radius.

6. Rust stains caused by reactive iron pyrites or other contaminants will occur where such contaminants are found as part of the aggregates. Rust stains may also be caused by particles of steel left by the aggregate crusher, pieces of tie wire from the cage assembly, or particles of steel burned off in welding and accidentally left in the mold. These stains (and steel particles) should be removed from the surface as soon as they are observed. Rust stains caused by corroding reinforcing steel are not common. When reinforcing steel does corrode, it reflects shortcomings in design, concrete quality or workmanship. Rust stains due to corrosion of hardware should not occur if the hardware has been protectively coated or where it is entirely behind a weatherproofed joint.

7. It should be recognized that some blemishes or variations in color occur in architectural precast concrete panels. Uniformity in color is directly related to ingredients supplying the color.

Panels containing aggregates and matrices of contrasting colors will appear less uniform than those containing materials of similar color (as the size of the coarse aggregate decreases, less matrix is seen and the more uniform the color of the panel will appear). It is advisable to match the color or tone of the matrix to that of the coarse aggregate so minor segregation of the aggregate will not be noticeable.

Color uniformity is difficult to achieve on gray, buff, and pigmented concrete surfaces. The use of white cement will give better color uniformity than gray cement. Al-

Standard

8. Blocking stains evident on exposed surface.

9. Areas of backup concrete bleeding through the facing concrete.

10. Foreign material embedded in the face.

11. Visible repairs at 20 ft. (6 m) viewing distance.

12. Reinforcement shadow lines.

13. Cracks visible at a 20 ft (6 m) viewing distance.

2.10 Sealers or Clear Surface Coatings

If sealers or clear surface coatings are specified, they shall be tested on reasonably sized samples of

Commentary

lowable color variation in the gray cement is enough to cause noticeable color differences in precast concrete panels. The slightest change of color is readily apparent on the uninterrupted surfaces of smooth off-the-mold concrete, and any variation is likely to be regarded as a surface blemish. As a general rule, a textured surface is aesthetically more satisfactory (greater uniformity) than a smooth surface. The surface highlights and natural variations in aggregate color will, to a large extent, camouflage subtle differences in texture and color of the concrete. The degree of uniformity (different shadings and to some extent, depth of color) between panels and within panels in a sandblasted finish, as in all exposed aggregate processes is generally in direct proportion to the depth of exposure. For example, a light sandblasting may look acceptable on a small sample, but uniformity is rather difficult to achieve in reality.

Sunlight, and exposure to the elements may even out the variation to a great extent.

8. Blocking used to separate production pieces from each other in the storage yard or during shipment should consist of non-staining material. Blocking used for extended periods of time should allow the precast concrete unit to cure in a similar environment as the rest of the unit, both under and around the blocking, by not trapping moisture or preventing air circulation to the blocked area. Plastic bubble type pads are available and are well suited for this purpose. Lumber or padding wrapped with plastic should not be used for blocking, unless in an area that is not visible in the final structure.

12. Reinforcing steel in some finishes may show up as light shadow lines usually directly over the steel depending on mix, concrete cover, vibration of reinforcement, placing, etc. In a few cases, a dark shadow pattern is displaced from the steel above.

13. It should be recognized that a certain amount of crazing or cracking may occur without being detrimental. With respect to acceptability of cracks, the cause should be determined as well as the stress condition a crack will be under with the precast concrete unit in place.

While some of these cracks may be repaired and effectively sealed, their acceptability should be governed by the importance and the function of the panel under consideration. The decision regarding acceptability must be made on an engineering basis as well as on visual appearance.

C2.10 Sealers or Clear Surface Coatings

Sealers or clear surface coatings may be considered for the possible improvement of weathering characteristics. The quality of concrete normally specified for architectural precast con-

Standard

varying age, and their performance verified over a suitable period of exposure or be based on prior experience under similar exposure conditions. Sealers shall be applied in accordance with manufacturer's written recommendations. Any clear sealer used shall be guaranteed by the supplier or applicator not to stain, soil, darken or discolor the finish, cause joint sealants to stain the panel surface or affect the bond of the sealant. The manufacturers of both the sealant and the sealer shall be consulted before application, or the materials specified shall be pretested before application.

Commentary

crete, even with minimum practical thickness, does not require sealers for waterproofing.

DIVISION 3 – RAW MATERIALS AND ACCESSORIES

Standard

3.1 Concrete Materials

3.1.1 General

An inspector shall continually check for any change in materials or proportions that will affect the surface appearance.

3.1.2 Cement

The type and kind of cements shall be selected to provide predictable strength and durability as well as proper color. Cements shall conform to ASTM C150. Concrete mixes using cements conforming to ASTM C595, C845 or C1157 shall be tested and evaluated for the intended applications.

To minimize the color variation of the surfaces exposed to view in the finished structure, cement of the same type, brand and color from the same mill shall be used throughout a given project. The cement used in the work shall correspond to that upon which the selection of concrete proportions was based.

3.1.3 Facing Aggregates

Commentary

C3.1 Concrete Materials

C3.1.1 General

A change in aggregate proportions, color or gradation will affect the uniformity of the finish, particularly where the aggregate is exposed. In smooth concrete the color of the cement (plus pigment) is dominant. If the concrete surface is progressively removed by sandblasting, retarders or other means, the color becomes increasingly dependent on the fine and coarse aggregates.

C3.1.2 Cement

Unless otherwise specified, the producer should have the choice of type and kind of cement to use to achieve the specified physical properties. Different cements have different color and strength development characteristics that affect the desired concrete. Copies of the cement strength uniformity tests conducted in accordance with ASTM C917 should be requested from the cement supplier. The cement color exerts a considerable influence on the color of the finished product due to its tremendous surface area per unit of weight.

Colored cements conforming to ASTM C150 which are produced by adding pigments to white cement during the production process may also be used.

Cement performance can be influenced by atmospheric conditions, and cement has an influence on finishing techniques, mix design requirements and casting procedures. Normal production variables such as changes in water content, curing cycles, temperature, humidity and exposure to climatic conditions at varying strength levels all tend to cause color variation. Color variation in a gray cement matrix is generally greater than those matrices made with white cement. A gray color may be produced by using white cement with a black pigment or a blend of white and gray cement. Uniformity normally increases with increasing percentage of white, but the gray color is dominant.

The temperatures of the mixing water and aggregates play a more important role in determining the concrete temperature.

C3.1.3 Facing Aggregates

The choice of fine and coarse aggregates to be used for face mixes should be based on a visual inspection of samples prepared by the precaster.

Selection of aggregates should be governed by the following:

1. Aggregates should have proper durability and be free of staining or deleterious materials. They should be nonreactive with cement and available in particle shapes (rounded or cubical rather than slivers) required for good concrete and appearance.

2. Final selection of colors should be made from concrete samples that have the proper matrix and are finished in the same manner as planned for production. Some finish-

Standard

Fine Aggregate. Fine aggregates for face mixes, other than lightweight aggregates, shall consist of high quality natural sand or sand manufactured from coarse aggregate. Fine aggregates shall comply with ASTM C33, except for gradation which can deviate to achieve desired texture. Variations in fineness modulus of fine aggregate shall not exceed ± 0.20 from the value used for the mix design and the amount retained on any two consecutive sieves shall not change by more than 10 percent by weight of the total fine aggregate sample.

Fine aggregates shall be obtained from sources from which representative samples have been subjected to all tests prescribed in the governing specifications and the concrete-making properties of the aggregates have been demonstrated by trial mixes.

Coarse Aggregate. Coarse aggregates for face mixes other than lightweight aggregates shall conform to the requirements of ASTM C33, except for gradation.

The nominal maximum size of coarse aggregate in the face mix shall not exceed:

1. One-fifth of the narrowest dimension between sides of molds.
2. One-third of the thickness of panels.
3. Three-fourths of the minimum clear depth of cover.
4. Two-thirds of the spacing between individual

Commentary

ing processes change the appearance of the aggregates. If small concrete samples are used to select the aggregate color, the architect/engineer should be aware that the general appearance of large areas after installation tends to be different than indicated by the trial samples.

3. Aggregates with a dull appearance may appear brighter in a white matrix than a gray matrix.

4. Weathering may influence newly crushed aggregate. When first crushed, many aggregates are bright but will dull slightly with time. Similarly, some of the sparkle caused by acid etching or bushhammering may not survive more than a few weeks. The architect/engineer should recognize that samples maintained indoors may not retain their exact appearance after exposure to weather for a few weeks.

5. The method used to expose the aggregate in the finished product may influence the final appearance.

6. The maximum size of coarse aggregate is usually controlled by (1) the dimensions of the unit to be cast, (2) clear distance between reinforcement, (3) clear distance between the reinforcement and the form, and (4) the desired finish.

Fine Aggregate. Fine aggregates have a major effect on the color of white and light buff colored concrete, and can be used to add color tones. Where the color depends mainly on the fine aggregates, gradation control is required, particularly where the color tone depends on the finer particles.

For the fine aggregate, the material passing the No. 100 (150-μm) sieve should not exceed 5 percent, and the maximum variation of the material passing the No. 100 (150-μm) sieve from the fine aggregate used in the initial mix design should not exceed 1 percent to ensure uniformity of concrete mixes.

Coarse Aggregate. Coarse aggregates may be selected on the basis of color, hardness, size, shape, gradation, method of surface exposure, cost and availability provided levels of strength, durability and workability are met. Colors of natural aggregates may vary considerably according to their geological classification and even among rocks of one type.

Aggregate size should also be selected on the basis of the total area to be cast and the distance from which it is to be viewed. Aggregates exposed on the face of the precast concrete unit may vary from 1/4 in. (6 mm) up to stones and rubble 6 to 7 in. (150 to 175 mm) in diameter and larger. Larger aggregates are required on large areas for any degree of apparent relief. When surfaces are some distance from the main flow of traffic, large aggregate is required for a rough-textured look. A suggested visibility scale is given in Table C3.1.3a.

Standard

reinforcing bars or bundles of bars or pretensioning tendons or post-tensioning ducts.

5. The minimum rib size, unless workability and consolidation methods are such that the concrete can be placed without honeycomb or voids.

Coarse aggregates shall be obtained from sources from which representative samples have been subjected to all tests prescribed in the governing specifications and the concrete-making properties have been demonstrated.

Once a sample panel has been approved by the architect/engineer, no other source of exposed aggregate or facing material shall be used for the project unless shown to be equivalent in quality, gradation and color to the approved sample.

The precast concrete manufacturer shall verify that an adequate supply from one source (pit or quarry) for each type of aggregate for the entire job will be readily available and, if possible, obtain the entire aggregate supply prior to starting the project or have the aggregate supply held by the supplier.

Facings of any suitable material such as natural stone, thin brick, ceramic tile, terra-cotta, oversized natural or crushed aggregates, aluminum or stainless steel sheets or sections may also be used as facing materials. Each of these special facing applications shall be properly designed and tested before use both with respect to suitability of the material and to the effect of its inter-relationship with the precast concrete.

When an aggregate source is specified that does not meet the requirements of this Manual, the precaster shall notify the architect/engineer in writing before the start of production.

3.1.4 Backup Aggregates

Aggregates in backup concrete shall comply with ASTM C33 or ASTM C330. In general, the maximum size of coarse aggregate shall not exceed:

1. One-third of the thickness of panels
2. Three-fourths of the minimum clear depth of cover.
3. Two-thirds of the spacing between individual reinforcing bars or bundles of bars or pretensioning tendons or post-tensioning ducts.

All backup aggregates shall be from approved sources from which representative samples have been subjected to all tests prescribed in the govern-

Commentary

Table C3.1.3a. Suggested visibility scale.

Aggregate size, in. (mm)	Distance at which texture is visible, ft (m)
1/4 – 1/2 (6-13)	20 – 30 (6-9)
1/2 – 1 (13-25)	30 – 75 (9-23)
1 – 2 (25-50)	75 – 125 (23-38)
2 – 3 (50-75)	125 – 175 (38-53)

Stockpiling of aggregates for an entire project will minimize color variation caused by variability of material and will maximize color uniformity.

Standard

ing specifications and the concrete making properties have been satisfactorily demonstrated.

3.1.5 Aggregates for Lightweight Concrete

Lightweight aggregates shall conform to the requirements of ASTM C330. Provisions for testing shall be as stipulated in Articles 6.1.2 and 6.1.3 except tests for gradation, unit weight, and impurities shall be made in accordance with requirements of ASTM C330.

3.1.6 Mixing Water

Water shall be free from deleterious matter that may interfere with the color, setting or strength of the concrete.

Water, either potable or non-potable, shall be free from injurious amounts of oils, acids, alkalies, salts, organic materials, chloride ions or other substances that may be deleterious to concrete or steel. The water shall not contain iron or iron oxides which will

Commentary

C3.1.5 Aggregates for Lightweight Concrete

Precasters using lightweight aggregates should be experienced in mixing and placing lightweight concrete mixes since their weight and shrinkage characteristics often require special attention in order to obtain a reasonable uniformity in appearance when exposed. The combination of normal weight face mix and a backup mix with lightweight aggregates may increase the possibility of bowing or warping. Before producing such a combination, pilot units, produced and stored under anticipated production conditions, are desirable to verify satisfactory performance.

Lightweight aggregates tend to take on moisture and if not saturated will pull water from the mix causing a rapid slump loss creating problems in handling and placing. The ACI Committee 213 report, Guide for Structural Lightweight Aggregate Concrete, provides a thorough discussion of lightweight aggregate properties including proportioning and mixing practices.

C3.1.6 Mixing Water

Excessive impurities in mixing water not only may affect setting time and strength, but also may cause efflorescence, staining, increased volume change and reduced durability. Therefore, certain optional limits should be set on chlorides, sulfates, alkalis, and solids in the mixing water or appropriate

Table 3.1.6. Chemical limits for non-potable mixing water.

Substance	Maximum Concentration, ppm	ASTM Test Method[a]
Soluble carbon dioxide	600	D513
Calcium plus magnesium	400	D511
Chloride, as Cl	500	D512
Iron	20	b
Phosphate	100	D4327
Arsenic	100	D2972
Boron and borates	100	D3082
Alkalies, as $Na_2O + 0.658 K_2O$	600	D4191 & 4192
Silt or suspended particles	2000	D1888
pH	6.0 to 8.0	D1067
Dissolved solids	2000	D1888
Iodate	500	c
Sugars and oils	Not detectable	c

a Other methods as used by water analysis companies are generally satisfactory.
b Alternatively, soluble carbonate and bicarbonate may be determined by appropriate methods (e.g., Hach Chemical Company procedures) and calculated as CO_2).
c No ASTM method available.

Standard

cause staining in light colored or white concrete. Water from a source other than a municipal water supply shall be tested on an annual basis as required in Article 6.1.2. The water shall not exceed the maximum concentration limits given in Table 3.1.6.

3.1.7 Admixtures

If a satisfactory history of admixture performance with the concrete making materials to be used in a project is not available, a trial mixture program with those materials, particularly the cement, shall be conducted. The trial mixture program shall demonstrate satisfactory performance of the admixture relative to slump, workability, air content, and strength under the conditions of use, particularly with respect to temperature and humidity. Admixtures shall be carefully checked for compatibility with the cement or other admixtures used to ensure that each performs as required without affecting the performance of the other admixtures. Admixture supplier's recommendations shall be observed subject to plant checking and experience. The effect of variations in dosage and the sequence of charging the admixtures into the mixer shall be determined from the recommendations of the admixture supplier or by trial mixes.

The same brand and type admixtures shall be used throughout any part of a project where color uniformity is required.

Air entraining admixtures shall conform to the requirements of ASTM C260.

Water reducing, retarding or accelerating admixtures shall conform to the requirements of ASTM C494. High-range water-reducing admixtures (HRWRA) or (superplasticizers) shall conform to the requirements of ASTM C494 Type F or G, or for flowing concrete to ASTM C1017, Type 1 or 2. Admixtures containing chloride ions shall not be used in prestressed concrete, or in concrete containing aluminum embedments or galvanized reinforcement and/or hardware, if their use will produce a deleterious concentration of chloride ions in the mixing water and cause corrosion.

Commentary

tests can be performed to determine the effect the impurity has on various properties. Some impurities may have little effect on strength and setting time, yet they can adversely affect durability and other properties.

The chloride ion content should be limited to a level well below the recommended maximum, if practical. Chloride ions contained in the aggregate and in admixtures should be considered in evaluating the acceptability of total chloride ion content of mixing water.

C3.1.7 Admixtures

All types of admixtures used should be materials of standard manufacture having well established records of tests to confirm their properties. Expected performance of a given brand, class, or type of admixture may be projected from one or more of the following sources:

1. Results from jobs which have used the admixture under good technical control, preferably using the same materials and under conditions similar to those to be expected.

2. Technical literature and information from the manufacturer of the admixture.

3. Laboratory tests made to evaluate the admixture.

Trial mixtures can be made at midrange slump and air contents expected or specified for the project. The cement content or water/cement ratio should be that required for the specified design strength and durability requirements for the job. Trial mixtures also can be made with a range of cement contents — water/cement ratios, slumps or other properties to bracket the project requirements. In this manner, the optimum mixture proportions can be selected and the required results achieved.

Various results can be expected with a given admixture due to differences in dosage, cement composition and fineness, cement content, aggregate size and gradation, the presence of other admixtures, addition sequence, changes in water/cement ratio, and weather conditions from day to day.

Differences in setting times and early strength development also can be expected with different types and sources of cement as well as concrete and ambient temperatures.

The use of air entrainment is recommended to enhance durability when concrete will be subjected to freezing and thawing while wet.

Use of water-reducing admixtures results in a desirable reduction in water-cement ratio for a given consistency (slump) and cement content, an increased consistency for the same water-cement ratio and cement content or obtaining specified strength at lower cement content while keeping the water-cement ratio the same. The reduction in water-cement ratio achieved by eliminating excess mixing water may produce greater strength improvement than a similar reduction obtained by adding cement. Generally, the effect of use of these materials on the hardened concrete is increased compressive strength and some reduction in permeability and, in combination with adequate air entrainment, improved resistance to freezing and thawing. Water-reducing admixtures also may improve the properties of concrete containing harsh facing mixes which necessarily have

Standard

To avoid corrosion problems admixtures containing chloride ions shall be limited to a maximum water soluble chloride ion (Cl⁻) in prestressed concrete to 0.06 percent by weight of cement or 0.30 in reinforced concrete when tested in conformance to ASTM C1218.

Mineral admixtures or pozzolans meeting ASTM C618 or C1240 may be added for additional workability, increased strength and reduced permeability and efflorescence provided no detrimental change is experienced in the desired architectural appearance. If a HRWRA is to be used with silica fume, ensure that the admixture to be used is compatible with that already in the silica fume, if any. The amount of silica fume or metakaolin in concrete shall not exceed 10 percent by mass of the portland cement unless evidence is available indicating that the concrete produced with a larger amount will have satisfactory strength, durability, and volume stability.

Coloring admixtures or pigments shall conform to the requirements of ASTM C979. All coloring pigments required for a project shall be ordered in one lot. The coloring pigment shall be a finely ground natural or synthetic mineral oxide or an organic phthalocyanine dye with a history of satisfactory color stability in concrete. Pigments shall be insoluble in water, free of soluble salts and acids, colorfast in sunlight, resistant to alkalies and weak acids, and virtually free of calcium sulfate. The amount and type of pigment used shall be harmless to concrete setting time or strength. Amounts of pigment used shall not exceed 10 percent of the weight of cement.

Commentary

poor aggregate gradation. Dosages required to produce specific results are usually recommended by the manufacturers.

There are mid-range water-reducing admixtures which may be classified under ASTM C494 as Type A or Type F depending on dosage rate.

Retarding admixtures are used primarily to offset the accelerating and damaging effect of high temperature, or in some cases to keep concrete plastic for a sufficiently long period of time (retard or control the initial set of the concrete) so that succeeding lifts can be placed without development of cold joints or discontinuities in the unit. For example, to achieve good bond between facing and backup concrete.

High-range water-reducing admixtures can be used to significantly increase slump without adding more water, or to greatly reduce water content without a loss in slump. Concretes containing these admixtures, particularly those with initial slumps less than 3 to 4 in. (75 to 100 mm) and low water-cement ratios, tend to lose slump and stiffen rapidly. While, some high-range water-reducing retarders can maintain the necessary slump for extended periods at elevated concrete temperatures.

Calcium chloride and admixtures containing chloride ions will promote corrosion of steel reinforcement and galvanized or aluminum embedments, may cause non-uniformity in color of the concrete surface (darkening and mottling), and may disrupt the efficiency of surface retarders.

Where a particularly smooth surface is desired, the addition of fine minerals or pozzolans conforming to ASTM C618 may be desirable. These materials also may be added to improve workability or to reduce the possibility of efflorescence provided no detrimental change is experienced in the desired architectural appearance. The use of fly ash or silica fume (microsilica) in a concrete mixture will darken the concrete color and may make it difficult to achieve color uniformity. The color of silica fume depends on carbon content and several other variables. Silica fume from one source could be almost white in color, while that from another may be black. Metakaolin is a white dry powder and does not darken white or gray concrete.

Pigments often are added to the matrix to obtain colors which cannot be obtained through combinations of cement and fine aggregate alone. Variable amounts of a pigment, expressed as a percentage of the cement content by weight, produce various shades of color. High percentages of pigment reduce concrete strength because of the high percentage of fines introduced to the mix by the pigments. For these reasons, the amount of pigment should be controlled within the limits of strength and absorption requirements. Different shades of color can be obtained by varying the amount of coloring material or by combining two or more pigments. Brilliant concrete colors are not possible with either natural or synthetic pigments due to their low allowable addition rates and the masking effects of the cement and aggregates. White portland cement will pro-

Standard

3.2 Reinforcement and Hardware

3.2.1 Reinforcing Steel

Steel reinforcing bars shall be deformed bars of the designated types of steel, sizes and grades and shall conform to the following applicable specifications as shown on the production drawings:

Billet-Steel Deformed Bars	ASTM A615/A615M
Low Alloy Steel Deformed Bars	ASTM A706/A706M
Rail-Steel Deformed Bars	ASTM A616/A616M
Axle-Steel Deformed Bars	ASTM A617/A617M

It shall be permissible to substitute: a metric size bar of Grade 300 for the corresponding inch-pound size bar of Grade 40; a metric size bar of Grade 350 for the corresponding inch-pound size bar of

Commentary

duce cleaner, brighter colors and should be used in preference to gray cement.

When using pigment dosages of less than 1% by weight of cement, the sensitivity of color intensity to minor pigment quantity variations is very high, causing potential unit to unit color variation. When using dosages from 1% to 5%, this sensitivity is much lower, and color variation will be more easily controlled. Addition of synthetic iron oxide pigments above 5% will not increase color intensity, while for natural pigments the saturation points are closer to 10%.

Coloring pigments of iron oxides are generally preferred because of better performance, but they may react chemically with other products, such as surface retarders or muriatic acid, and should be tested prior to use.

Green is quite permanent, except in light shades. Some blues are not uniform or permanent. Cobalt blue should be used to avoid problems. Dark colors have more tendency to show efflorescence that forms on all concrete surfaces. If the lightening of the color becomes too objectionable, the color can be restored by washing with dilute hydrochloric acid and rinsing thoroughly. Carbon black, due to its extremely fine particle size, has a tendency to wash out of a concrete matrix and is not recommended. Synthetic black iron oxide will produce a more stable charcoal color.

Architects can best specify the color they desire by referring to a swatch or color card. A cement color card is preferable but one published by a paint manufacturer is acceptable. An excellent color reference is the Federal Color Standard 595 B, published by the U.S. Government Printing Office.

Efflorescence deposited on the surface may mask the true color and give the appearance of fading even though the cement paste itself has undergone no change. In addition, weathering of the pigmented cement paste exposes more of the aggregate to view. If the color of the aggregate is in contrast to that of the pigment, a change in the overall color of the surface may be noted.

C3.2 Reinforcement and Hardware

C3.2.1 Reinforcing Steel

Grades of reinforcing steel required for a specific application are determined by the structural design of the precast concrete units.

Many mills will mark and supply bars only with the metric designation, which is a soft conversion. Soft means that the metric bars have exactly the same dimensions and properties as the equivalent in.-lb designation.

The size of reinforcing bars is often governed by dimensions of the element, required concrete cover over steel, and function of the element. In general, bar sizes should be kept reasonably small even where this will reduce the spacing of the bars. Smaller bars closely spaced will decrease the size of potential cracks and improve the distribution of temperature stresses. The use of additional reinforcing bars as compared with fewer heavier bars becomes more important in thinner concrete sections. Since the sum of potential cracks in concrete is more or

Standard

Grade 50; a metric size bar of Grade 420 for the corresponding inch-pound size bar of Grade 60; and a metric size bar of Grade 520 for the corresponding inch-pound size bar of Grade 75.

Zinc-coated (galvanized) reinforcement shall conform to ASTM A767/A767M and be chromate treated.

Epoxy coated reinforcement shall conform to ASTM A775/A775M or A934/A934M. Any plant supplying epoxy coated reinforcement shall be a participant in the CRSI Voluntary Certification Program for Fusion-Bonded Epoxy Coating Applicator Plants.

Fading of the epoxy coating color shall not be cause for rejection of epoxy coated reinforcing bars.

Bar mats shall conform to ASTM A184/A184M and shall be assembled from the bars described above. If bars other than the types listed above are to be used, their required properties shall be shown on the production drawings. In addition to the ASTM specification requirements, all reinforcing bars shall meet the requirements of ACI 318.

The weldability of reinforcing bars other than ASTM A706/A706M shall be evaluated according to provisions of AWS D1.4.

Welded wire reinforcement shall conform to the following applicable specifications:

Plain Wire	ASTM A82
Deformed Wire	ASTM A496
Welded Plain Wire Reinforcement	ASTM A185
Welded Deformed Wire Reinforcement	ASTM A497

Galvanized welded wire reinforcement shall be made from zinc-coated (galvanized) carbon steel wire conforming to ASTM A641; or be hot-dipped galvanized and be chromate treated; or be allowed to weather. Epoxy-coated welded wire reinforcement shall conform to ASTM A884/A884M, Class A. All damaged areas of epoxy coating shall be repaired (touched-up) with patching material.

Welded wire reinforcement mesh spacings and wire

Commentary

less constant for a given set of conditions, the more bars there are the smaller and less visible the cracks.

Where galvanizing of reinforcing bars is required, galvanizing is usually performed after fabrication. The ASTM A767/A767M specification prescribes minimum finished bend diameters for bars that are fabricated before galvanizing. Smaller finished bend diameters are permitted if the bars are stress-relieved. The ASTM A767/A767M specification has two classes of zinc coating weights. Class II [2.0 oz./sq. ft (610 g/m^2)] is normally specified for precast concrete units.

When epoxy-coated reinforcing bars are exposed to sunlight over a period of time, fading of the color of some epoxy coatings may occur. Since the discoloration does not harm the coating nor affect its corrosion-protection properties, such fading should not be cause for rejection of the coated bars.

ASTM A706/A706M is specifically intended for welding. Chemical analyses are not ordinarily meaningful for rail-steel (ASTM A616/A616M) and axle-steel (ASTM A617/A617M) reinforcing bars. Welding of these bars is not recommended.

Welded wire reinforcement may be used as the main reinforcement and, if necessary, reinforcing bars are added in ribs or other locations to provide the area of steel required.

Galvanized or epoxy-coated reinforcement is generally required when cover over reinforcement is 3/4 in. (19 mm). In these cases, the use of galvanizing or epoxy-coating should be specifically called for in the contract documents and shown on the shop drawings.

It is recommended that welded wire reinforcement be pur-

Standard

sizes (gages) shall be shown on the production drawings. In addition to the ASTM specification requirements, all wire reinforcement shall meet the requirements of ACI 318.

Reinforcement, with rust, seams, surface irregularities, or mill scale shall be considered as satisfactory, provided the minimum nominal dimensions, including minimum average height of deformations, and nominal weight of a hand-wire-brushed test specimen are not less than the applicable ASTM specification requirements.

Plastics for welded wire and bar supports shall be composed of polyethylene, styrene copolymer rubber-resin blends, poly-(vinyl chlorides), Types I and II, and polytetrafluoroethylene. Plastics shall be alkali resistant and should have at least 25 percent of their gross plane area perforated.

3.2.2 Prestressing Materials

Strand materials for prestressing shall consist of:

1. Pretensioning

    a. Uncoated, low-relaxation strand conforming to ASTM A416, Grade 250 (1725) or Grade 270 (1860).

    b. Uncoated, stress-relieved (normal relaxation) strand, conforming to ASTM A416, Grade 250 (1725) or Grade 270 (1860).

2. Post-Tensioning

    a. Strand as described above either singly or in multiple parallel strand units with wedge type or other adequate anchorages.

    b. Uncoated, stress-relieved wire conforming to ASTM A421 in multiple parallel wire units with wedge-type, button head or other adequate anchorages.

    c. High strength, stress-relieved bars conforming to ASTM A722 with wedge type, threaded, or other adequate anchorages.

A light bond coating of tight surface rust on prestressing tendons is permissible, provided strand surface shows no pits visible to the unaided eye after rust is removed with a non-metallic pad.

Commentary

chased in large sheets rather than rolls for better control of flatness.

Plastic bar supports and spacers have about fifteen times the thermal expansion coefficient of concrete. The plane surface has to be perforated to permit the concrete to weave into the section and restrain movement. This inhibits thermal punching and transverse cracking in thin sections.

C3.2.2 Prestressing Materials

Due to the bond development required of concrete to prestressing strand, bars or wires, the surface condition of tendons is critical to prestressed concrete. The presence of light rust on a strand has proven to be an enhancement to bond over bright strand and therefore should not be a deterrent to the use of the strand. A pit visible to the unaided eye, when examined as described in "Evaluation of Degree of Rusting on Prestressed Concrete Strand," Sason, Augusto S., PCI JOURNAL, May-June 1992, V.37, No. 3, pp. 25-30 is cause for rejection. A pit of this magnitude is a stress raiser and greatly reduces the capacity of the strand to withstand repeated or fatigue loading. In

Standard

Strand chucks for pretensioning shall be capable of anchoring the strand without slippage after seating. Length of grips and configuration of serrations shall be such as to ensure against strand failure within the vise jaws at stresses less than 95 percent of strand ultimate strength. Steel casings for strand vises shall be verified by the manufacturer as capable of holding at least 100 percent of the ultimate strength of the strand.

Tendon anchorages for post-tensioning shall meet the following requirements:

1. Anchorages for bonded tendons tested in an unbonded state shall develop 95 percent of the actual ultimate strength of the prestressing steel, without exceeding anticipated set at time of anchorage. Anchorages which develop less than 100 percent of the minimum specified ultimate strength shall be used only where the bond length provided is equal to or greater than the bond length required to develop 100 percent of the minimum specified ultimate strength of the tendon. The required bond length between the anchorage and the zone where the full prestressing force is required under service and ultimate loads shall be sufficient to develop the specified ultimate strength of the prestressing steel. Determine the bond length by testing a full-sized tendon. If in the unbonded state the anchorage develops 100 percent of the minimum specified ultimate strength it need not be tested in the bonded state.

2. Anchorages for unbonded tendons shall develop 95 percent of the minimum specified ultimate strength of the prestressing steel with an amount of permanent deformation which will not decrease the expected ultimate strength of the assembly.

3. The minimum elongation of a strand under ultimate load in an anchorage assembly tested in the unbonded state shall be not less than 2 percent when measured in a gauge length of 10 ft (3 m).

Anchorage castings shall be nonporous and free of sand, blow holes, voids and other defects.

For wedge type anchorages, the wedge grippers shall be designed to preclude premature failure of

Commentary

many cases, a heavily rusted strand with relatively large pits will still test to an ultimate strength greater than specification requirements. However, it will not meet the fatigue test requirements.

Strand chuck maintenance should be in force for all elements in use based on guidelines in Article 5.3.5 and Appendix D.

Post-tensioning tendons subject to exposure or condensation, and which are not to be grouted, should be permanently protected against corrosion by plastic coating or other approved means.

3. Elongation based on tests with a gauge length less than 10 ft (3 m) should not be cause for rejection.

Standard

the prestressing steel due to notch or pinching effects under static test load conditions to determine yield strength, ultimate strength, and elongation of the tendon.

Anchorages other than the types listed may be used provided they are shown by an adequate program of tests to meet the basic requirements listed above.

Sheathing for bonded post-tensioned tendons shall be strong enough to retain its shape, resist unrepairable damage during production, and prevent the entrance of cement paste or water from the concrete. Sheathing material left in place shall not cause harmful electrolytic action or deteriorate. The inside diameter shall be at least 1/4 in. (6 mm) larger than the nominal diameter of single wire, bar, or strand tendons; or in the case of multiple wire, bar, or strand tendons, the inside cross-sectional area of the sheath shall be at least twice the net area of the prestressing steel. Sheaths shall be capable of transmitting forces from the grout to the surrounding concrete. Sheaths shall have grout holes or vents at each end and at all high points except where the degree of tendon curvature is small and the tendon is relatively level.

Grout shall consist of a mixture of cement and water unless the gross inside cross-sectional area of the sheath exceeds four times the tendon cross-sectional area, in which case fine aggregate may be added to the mixture. Fly ash and pozzolanic mineral admixtures may be added at a ratio not to exceed 0.30 by weight of cement. Mineral admixtures shall conform to ASTM C618. Aluminum powder of the proper fineness and quantity or other approved shrinkage-compensating material which is well dispersed through the other admixture may be used to obtain 5 to 10 percent unrestrained expansion of the grout. Admixtures containing more than trace amounts of chlorides, fluorides, aluminum, zinc, or nitrates shall not be used. Fine aggregate, if used, shall conform to ASTM C404, Size No. 2, except that all material shall pass the No. 16 sieve. Grout shall achieve a minimum compressive strength of 2500 psi (17.2 MPa) at 7 days and 5000 psi (34.5 MPa) at 28 days when tested in accordance with ASTM C1107, and have a consistency that will facilitate placement. Water content shall be the minimum necessary for proper placement, and the water-cement ratio shall not exceed 0.45 by weight.

Sheathing for unbonded tendons (monostrand post-tensioning system) shall be polypropylene, high-density polyethylene, or other plastic which is

Commentary

Different requirements are imposed upon sheathings for bonded and unbonded tendons. In unbonded tendons, the sheathing does not transmit any bond stresses from the prestressing steel to the concrete and therefore has to assure the freedom of movement of the prestressing steel and form an adequate cover over the coated tendon. In bonded tendons, bond stresses will be transmitted through the sheathing, and it must be of such material and/or configuration to effectively allow this stress transfer.

The void in the concrete in which the tendon is to be located can also be pre-formed (e.g. by inflatable and removable tubes) and the tendon subsequently pulled through. With pre-formed voids no additional sheathing will be required.

Due to variations in the manufacturing process, slight variations may occur concentrically in the wall thickness. The sheathing should provide a smooth circular outside surface and

Standard

not reactive with concrete, coating, or steel. The material shall be watertight and have sufficient strength and durability to resist damage and deterioration during fabrication, transport, storage, installation, concreting, and tensioning. The sheath shall have a coefficient of friction with the strand of less than 0.05. Tendon covering shall be continuous over the unbonded length of the tendon and shall prevent the intrusion of water or cement paste and the loss of the coating material during concrete placement. The sheaths shall not become brittle or soften over the anticipated exposure temperature and service life of the structure. The minimum wall thickness of sheaths for noncorrosive conditions shall be 0.04 in. (1 mm.) The sheathing shall have an inside diameter at least 0.030 in. (0.76 mm) greater than the maximum diameter of the strand.

Tendons shall be lubricated and protected against corrosion by a properly applied coating of grease or other approved material. Minimum weight of coating material on the prestressing strand shall be not less than 2.5 pounds (1.1 kg) of coating material per 100 ft (30.5m) of 0.5 in. (12mm) diameter strand, and 3.0 pounds (1.4 kg) of coating material per 100 ft (30.5m) of 0.6 in. (15.24mm) diameter strand. The amount of coating material used shall be sufficient to ensure essentially complete filling of the annular space between the strand and the sheathing. The coating shall extend over the entire tendon length. Coatings shall remain ductile and free from cracks at the lowest anticipated temperature and shall not flow out from the sheath at the maximum anticipated temperature. Coatings shall be chemically stable and nonreactive to the tendon, the concrete and the sheath.

3.2.3 Hardware and Miscellaneous Materials

All hardware — connection items, inserts or other appurtenances — shall be clearly detailed in the project documents showing size and yield strength for architect/engineer approval.

Hardware shall be made from materials that are ductile. Plates and angles shall be low carbon (mild) steel and steel for anchors shall be of a grade and strength similar to the hardware material which it anchors to minimize welding complications. Brittle materials, such as low shock resistant, high carbon steels or gray iron castings, shall not be used. Malleable cast iron is satisfactory.

Commentary

should not visibly reveal the lay of the strand.

C3.2.3 Hardware and Miscellaneous Materials

Precautions should be taken to assure that hardware elements welded together are compatible. The degree of protection from corrosion required will depend on the actual conditions to which the connections will be exposed in service. The most common condition requiring protection is exposure to climatic conditions. Connection hardware generally needs protection against humidity or a corrosive environment. Corrosion could cause subsequent rusting and marring of adjacent elements or failure of the unit or connection. The use of oil based primers containing lead may be restricted due to local environmental regulations. Protective coatings should be of such quality and applied in such a manner that embrittlement cannot occur. Also, no loss of connection strength or reinforcement bond loss

Standard

Materials used in ferrous items embedded in the concrete, for the purpose of connecting precast elements or attaching or accommodating adjacent materials or equipment, shall conform to the requirements of the following specifications:

Structural steel: ASTM A36/A36M, (for carbon steel connection assemblies) except that silicon (Si) content shall be in the range of 0 to 0.04% or 0.15 to 0.20% and phosphorus (P) content in the range of 0 to 0.02% for materials to be galvanized. Steel with chemistry conforming to the formula $Si + 2.5P \leq 0.09$ is also acceptable.

Stainless steel: ASTM A666, Type 300 series, Grades A or B, (stainless steel anchors for use when resistance to staining merits extra cost.)

Carbon steel plate: ASTM A283/A283M, Grades A, B, C or D.

Malleable iron castings: ASTM A47/A447M, Grades 32510 or 35028.

Carbon steel castings: ASTM A27/A27M, Grade 60-30 (for cast steel casting clamps).

Anchor bolts: ASTM A307 (carbon steel) or A325/A325M (high strength steel) for low-carbon steel bolts, nuts and washers).

Carbon steel bars: ASTM A675/A675M, Grade 65 (for completely encased anchors).

Carbon steel structural tubing: ASTM A500, Grade B (for rounds and shapes).

High-strength low-alloy structural steel: ASTM or A572/A572M except that silicon (Si) content shall be in the range of 0 to 0.04% or 0.15 to 0.20% and phosphorus (P) content in the range of 0 to 0.02% for materials to be galvanized. Steel with chemistry conforming to the formula $Si + 2.5P \leq 0.09$ is also acceptable.

Welded headed studs: ASTM A108 Grades 1010 through 1020 inclusive or ASTM A276 (stainless steel) with the following mechanical property requirements (Table 3.2.3):

Commentary

should occur which had not been anticipated and allowed for in the design. Often, final finishing of the products causes the protective finish of the hardware to be damaged. When this occurs, a final touch up coating of the original protective material is required. This work should be performed in accordance with the recommendations of the coating material manufacturer. Since the final connection of a unit to a structure may require a field weld, the protective coating (zinc rich or epoxy paint) should be applied according to the manufacturer's requirements after final welding and cleaning of the welded area.

Standard

Table 3.2.3. Minimum mechanical property requirements for studs.

Physical Property	Type A AWS D1.1	Type B AWS D1.1	Type C ASTM A496 (deformed bars – any size)	Stainless Steel ASTM A276 AWS D1.6
Tensile Strength, psi (MPa)	55,000 (380)	60,000 (415)	80,000 (552)	75,000 (517)
Yield Strength, psi (MPa)				
(0.2% off-set)	—	50,000 (345)		30,000 (207)
(0.5% off-set)	—	—	70,000 (485)	
Elongation (% in 2 in.)	17	20	—	30
Reduction of Area, %	50	50	—	40

All metallic hardware surfaces exposed to the weather in service or subject to corrosive conditions, including condensation, shall have the portions of the hardware within 1/2 in. (12mm) of the concrete surface protected against corrosion or be made of non-corrosive materials. Hardware shall be properly cleaned prior to application of protective treatment.

Corrosion protection, when required, shall consist of one of the following:

1. Shop primer paint — FS-TT-P-645 or 664, or SSPC-Paint 25.

2. Zinc-rich paint (95 percent pure zinc in dried film) — FS-TT-P-641, Type III, or DOD-P-21035, self-curing, one component, sacrificial organic coating or SSPC-Paint 20.

3. Zinc metalizing or plating — ASTM B633.

4. Cadmium plating — ASTM B766 (particularly appropriate for threaded fasteners).

5. Hot dip galvanizing — ASTM A123 or A153.

6. Epoxy coating.

7. Stainless steel.

Commentary

4. Cadmium coatings will satisfactorily protect steel embedded in concrete, even in the presence of moisture and normal chloride concentrations. Minor imperfections or breaks in the coating will generally not promote corrosion of the underlying steel.

7. Some designers have specified the use of stainless steel connections to prevent long-term corrosion. While this

Standard

8. Other coatings or steels proven suitable by test.

Threaded parts of bolts, nuts or plates shall not be hot dip galvanized or epoxy coated unless they are subsequently rethreaded prior to use. Connection hardware shall be galvanized, if required, following fabrication. To avoid possible strain-age embrittlement and hydrogen embrittlement, the practices given in ASTM A143 shall be adhered to. Malleable castings shall be heat-treated, prior to galvanizing by heating to 1250 deg. F (677 deg. C) and water quenching.

Care shall be taken to prevent chemicals, such as muriatic acid, from contacting the hardware and causing corrosion.

Commentary

may appear to be the ultimate in corrosion protection, users are cautioned that the welding of stainless steel produces more heat than convention welding. That, plus a higher coefficient of thermal expansion, can create adverse hardware expansion adjacent to the assembly being welded, thus causing cracking in the adjacent concrete and promoting accelerated long-term deterioration. When stainless steel connection plates are used, edges should be kept free from adjacent concrete to allow expansion during welding without spalling the concrete. Heat dissipation can also be facilitated by the use of a thicker plate. In addition, the 300 Series stainless steels are susceptible to stress corrosion cracking when the temperature is over 140 deg. F (60 deg. C) and chloride solutions are in contact with the material.

8. Embedded natural weathering steels generally do not perform well in concrete containing moisture and chloride. Weathering steels adjoining concrete may discharge rust and cause staining of concrete surfaces.

In order to ensure that the strengths of the various elements of a connection are not reduced by hot-dip galvanizing, several precautions are necessary. When items of a connection assembly require welding, such as anchor bars to plates, the following recommendations by the American Hot-Dip Galvanizers Association have been found to produce satisfactory results:

1. An uncoated electrode should be used whenever possible to prevent flux deposits.

2. If coated electrode is used, it should provide for "self-slagging" as recommended by welding equipment suppliers. All welding flux residues must be removed by wire brushing, flame cleaning, chipping, grinding, needle gun or abrasive blast cleaning. This is necessary because welding flux residues are chemically inert in the normal pickling solutions used by galvanizers; their existence will produce rough and incomplete zinc coverage.

3. A welding process such as metal-inert gas (MIG), tungsten-inert gas (TIG), or CO_2 shielded arc is recommended when possible since they produce essentially no slag.

Special care should be taken when galvanized assembles are used. Many parts of connection components are fabricated using cold rolled steel or cold working techniques, such as bending of anchor bars. Any form of cold working reduces the ductility of steel. Operations such as punching holes, notching, producing fillets of small radii, shearing and sharp bending may lead to strain-age embrittlement of susceptible steels, particularly those with high carbon content. The embrittlement may not be evident until after the work has been galvanized. This occurs because aging is relatively slow at ambient temperatures but is more rapid at the elevated temperature of the galvanizing bath.

Standard

The materials of a connection shall be selected and joined in a manner such that embrittlement of any part of the assembled connection will not occur. Nonferrous inserts shall have proven to be resistant to electrolytic action and alkali attack. Documentation shall be provided showing satisfactory results over a reasonable period of time. If more than one material is used in a connection, abutting materials shall be selected such that corrosion is not induced. Dissimilar metals shall not be embedded near or in direct contact with each other in moist or saturated concrete unless experience has shown that no detrimental chemical or electrochemical (galvanic) reactions will occur or surfaces are permanently protected against corrosion.

Wooden inserts in the concrete shall be sealed to minimize volume changes during concrete placing, curing, and freezing weather conditions.

3.2.4 Handling and Lifting Devices

Since lifting devices are subject to dynamic loads, handling and lifting devices shall be fabricated from ductile material. Reinforcing bars shall not be used as lifting devices. If smooth bars are required for lifting, ASTM A36 steel of a known steel grade bent to correct size and shape shall be used provided adequate embedment or mechanical anchorage exists. Each bar size and configuration shall be substantiated, by testing, to ensure it meets load and handling requirements. The diameter shall be such that localized failure will not occur by bearing on the lifting device. Coil rods and bolts shall not be welded when used in lifting operations

To avoid overstress in one lifting loop when using multiple loops, care shall be taken in the fabrication to ensure that all strands are bent the same to assure even distribution of load between loops. Strands undergo physical changes when bent into loops; therefore, care shall be exercised with multiple bending.

Shop drawings shall clearly define insert dimensions and location for fabrication and placement or refer to standard details. Corrosion protection shall be considered where such hardware is left in the units.

All lifting devices shall be capable of supporting the element in all positions planned during the course of manufacture, storage, delivery and erection.

Commentary

Nonferrous metals embedded in concrete may corrode in two ways: (1) by direct oxidation in strong alkaline solutions normally occurring in fresh concrete and mortar; or (2) by galvanic currents that occur when two dissimilar metals are in contact in the presence of an electrolyte, or when an alloy or metal is not perfectly homogeneous, or when different parts of a metal have been subjected to different heat treatments or mechanical stresses.

Aluminum suffers attack when embedded in concrete. Initially, when aluminum is placed in fresh concrete, a reaction occurs resulting in the formation of aluminum oxide and the evolution of hydrogen. The greater volume occupied by these oxidation products causes expansive pressures around the embedded metal and may lead to increased porosity of the surrounding concrete as well as cracking and/or spalling of the concrete.

A wood sealer should be applied to prevent moisture migration from concrete into the wood. The high volume change of lumber which occurs even with changes in atmospheric humidity may lead to cracking of the concrete. Also, the embedment of lumber in concrete has sometimes resulted in leaching of the wood resins by calcium hydroxide with subsequent deterioration.

C3.2.4 Handling and Lifting Devices

The most common lifting devices are prestressing strand or cable loops projecting from the concrete, coil threaded inserts, or proprietary devices.

Deformed reinforcing bars should not be used since the deformations result in stress concentrations from the shackle pin. Also, reinforcing bars may be hard grade or re-rolled rail steel with little ductility and low impact strength at cold temperatures. Therefore, sudden impact loads, such as those encountered during stripping and handling, may cause failure.

Standard

Safe loads for lifting inserts or devices shall be established by full-scale tests to failure performed by a licensed professional engineer or supplied by manufacturer of proprietary devices. Information also shall be supplied by manufacturer on use and installation of the devices to assure proper performance.

Connection hardware shall not be used for lifting or handling, unless carefully reviewed and approved by the precast engineer.

3.3 Insulation

Insulation types shall conform to ASTM standards for material production. These standards present quality control minimums for each product matrix and are as follows:

Expanded Polystyrene
ASTM C-578   Type I, II, VIII, IX, XI

Extruded Polystyrene
ASTM C-578   Type IV, V, VI, VII, X

Polyurethane
ASTM C-591   Type 1, 2, 3

Polyisocyanurate
ASTM C-591   Type 1, 2, 3

Phenolic
ASTM C-1126   Type I, II, III

All relevant information on properties of insulating materials shall be on file at the plant.

Care shall be exercised when the insulation is exposed to temperatures greater than 140 deg. F (60 deg. C).

3.4 Welding Electrodes

Electrodes for shielded metal arc welding (SMAW) shall conform to the requirements of AWS D1.1, Section 4 or AWS D1.4, Section 5 (AWS A5.1 or A5.5). All welding electrodes shall be of a type suit-

Commentary

C3.3 Insulation

Cellular (rigid) insulations used in the manufacturing of sandwich panels come in two primary forms, thermoplastic and thermosetting. The thermoplastic insulations are known as molded expanded polystyrene or beadboard and extruded polystyrene or extruded board. Thermosetting insulations consist of polyurethane, polyisocyanurate and phenolic.

Although there are many insulation types on the market today, sandwich panels utilize a cellular (rigid) insulation due to the material properties needed to perform between two independent layers of concrete. These material properties include, but are not limited to, thermal and vapor transmission characteristics, moisture absorption, dimensional stability, coefficient of expansion and compressive and flexural strengths.

Since the insulation is generally placed in direct contact with the plastic concrete, excessive dewatering of the fresh concrete may occur with the use of highly absorbent insulation preventing cement from hydrating properly. The insulating quality of a material also will diminish if it absorbs moisture. For this reason, a material which exhibits no capillarity, or has suitable vapor barriers, should be selected. Some insulating materials such as molded polystyrene have high absorption.

C3.4 Welding Electrodes

Standard

Table 3.4.1. Filler metal requirements

Base Metal	Electrode Classifications for Welding Processes	
	Shielded Metal-Arc Low Hydrogen Electrodes	Flux-Cored Arc
ASTM A615 Grade 40 (300) ASTM A617 Grade 40 (300)	AWS A5.1 or A5.5 E70XX*	AWS A5.20 E7XT-X (Except -2, -3, -10, -GS)
ASTM A616 Grade 50 (350) ASTM A706 Grade 60 (420)	AWS A5.5 E80XX-X	AWS A5.29 E8XTX-X
ASTM A615 Grade 60 (420) ASTM A616 Grade 60 (420) ASTM A617 Grade 60 (420)	AWS A5.5 E90XX-X	AWS A5.29 E9XTX-X
ASTM A615 Grade 75 (520)	AWS A5.5 E100XX-X	AWS A5.29 E10XTX-X
ASTM A36 ASTM A500 Grade A Grade B	AWS A5.1 or A5.5 E60XX E70XX E70XX-X	AWS A5.20 E6XT-X E7XT-X (Except -2, -3, -10, -GS)
ASTM A441 ASTM A572 Grade 42 Grade 50	AWS A5.1 or A5.5 E7015, E7016 E7018, E7028 E7015-X, E7016-X E7018-X	AWS A5.20 E7XT-X (Except -2, -3, -10, -GS)
ASTM A572 Grade 60 Grade 65	AWS A5.5 E8015-X, E8016-X E8018-X	AWS A5.29 E8XTX-X

* XX = 15 or 16.

All low hydrogen electrodes conforming to AWS A5.1 shall be purchased in hermetically-sealed containers or shall be dried for at least 2 hrs. at a temperature between 450°F (230°C) and 500°F (260°C) before they are used.

All low hydrogen electrodes conforming to AWS A5.5 shall be purchased in hermetically-sealed containers or shall be dried at least 1 hr. at temperatures between 700°F (370°C) and 800°F (430°C) before being used.

able for the chemistry of the steel being welded (see Table 3.4.1).

The electrodes and the shielding for gas metal arc welding (GMAW) or flux-cored arc welding (FCAW), for producing weld metal with minimum specified yield strengths of 60 ksi (415 MPa) or less, shall conform to the requirements of AWS A5.18 or AWS A5.20.

Electrodes for welded reinforcing bar lap splices in contact or lap welds with splice plates or angles need not comply with Table 3.4.1. The length and size of welds shall be as shown on the shop drawings.

Weld metal having a minimum specified yield strength greater than 60 ksi (415 MPa) shall con-

Commentary

Design tensile capacity of lap welds is determined by length and size of welds using various electrodes not by compatibility of material tensile strength. However, the chemical composition of the materials should be compatible.

Standard

form to the following requirements:

1. The electrodes and shielding for gas metal arc welding for producing weld metal with a minimum specified yield strength greater than 60 ksi (415 MPa) shall conform with AWS A5.28.

2. The electrodes and shielding gas for flux-cored arc welding for producing weld metal with a minimum specified yield strength greater than 60 ksi (415 MPa) shall conform to AWS A5.29.

3. The plant shall have on file the electrode manufacturer's certification that the electrode will meet the above requirements of classification.

When a gas or gas mixture is used for shielding in gas metal arc or flux-cored arc welding, it shall be of a welding grade having a dew point of -40 deg. F (-40 deg. C) or lower. The plant shall have on file the gas manufacturer's certification that the gas or gas mixture will meet the dew point requirement.

All low hydrogen electrodes conforming to AWS A5.1 shall be purchased in hermetically-sealed containers or shall be dried for at least 2 hrs. at a temperature between 500 deg. F (260 deg. C) and 800 deg. F (430 deg. C) before they are used.

All low hydrogen electrodes conforming to AWS A5.5 shall be purchased in hermetically-sealed containers or shall be dried at least 1 hr. at temperatures between 700 deg. F (370 deg. C) and 800 deg. F (430 deg. C) before being used.

Electrodes shall be dried as specified above prior to use if the hermetically-sealed containers are damaged, improperly stored, or for any reason the electrodes are exposed to high moisture conditions. Immediately after removal from hermetically-sealed containers or from drying ovens, electrodes shall be stored in holding ovens and held at a temperature of 250 deg. F (120 deg. C) above ambient temperature. E70 series electrodes that are not used within 4 hrs., E80 series within 2 hrs., E90 series electrodes within 1 hr., or E100 series within 1/2 hr., after removal from hermetically-sealed containers or from a drying or storage oven, shall be redried in a drying oven before use. Electrodes shall be dried only once for any reason. Electrodes which have been wet shall not be used.

When joining different grades of steels, the electrode shall be selected for the lower strength base metal.

When steel cannot be completely cleaned of scale, rust, paint, moisture or dirt, an E6010 or E6011 electrode shall be used.

Commentary

All low hydrogen and stainless steel shielded metal arc electrode coverings should be protected from moisture pickup. Normally, electrodes packaged in hermetically sealed containers can be stored for several months without deteriorating. However, after the container is opened, the coating begins to absorb moisture and, depending on the ambient air condition, may need to be reconditioned after only four hours of exposure, otherwise porosity may result, especially at arc starts.

Only low hydrogen welding rods (EXX-X5, 6, or 8) should be kept in an oven once removed from their air tight container. Although they must not get wet, the coating of other rods (60-11) will be damaged if heated.

Standard	Commentary

Martensitic stainless steels that are to be postheated — (annealed, stress-relieved, or used under high temperature conditions) shall be welded with straight-chromium stainless steel electrodes of the E-400 series. Any of these alloys that will not be postheated shall be welded with austenitic chromium-nickel electrodes of the E-300 series.

The maximum diameter of electrodes shall be:

1. 5/16 in. (8.0 mm) for all welds made in the flat position, except root passes

2. 1/4 in. (6.4 mm) for horizontal fillet welds

3. 1/4 in. (6.4 mm) for root passes of fillet welds made in the flat position and groove welds made in the flat position with backing and with a root opening of 1/4 in. or more

4. 5/32 in. (4.0 mm) for welds made with low hydrogen electrodes in the vertical and overhead positions

5. 3/16 in. (4.8 mm) for root passes of groove welds and for all other welds not included above.

DIVISION 4 — CONCRETE

Standard

4.1 Mix Proportioning

The properties of concrete mixtures and the color and texture of the unit shall be as specified in the project specifications.

4.1.1 Qualification of New Concrete Mixes

Concrete mixes for precast concrete shall be established initially by laboratory methods. The proportioning of mixes shall be done either by a qualified commercial laboratory or qualified precast concrete quality control personnel. Mixes shall be evaluated by trial batches prepared in accordance with ASTM C192 and production tests under conditions simulating as closely as possible actual production and finishing. Tests shall be made on all mixes to be used in production of units. When accelerated curing is to be used, it is necessary to base the mix proportions on similarly cured test specimens.

Each concrete mix used shall be developed using the brand and type of cement, the source and gradation of aggregates, and the brand of admixture proposed for use in the production mixes. If any of these variables are changed the proportions of the mixture shall be re-evaluated.

Commentary

C4.1 Mix Proportioning

Much of the skill, knowledge and technique of producing quality architectural precast concrete centers around the proper proportioning of the mix. Before a concrete mix can be properly proportioned, several factors must be known. The finish, size and shapes of units to be cast should be considered. The method of consolidation should be known to determine the required workability. The maximum size of the coarse aggregate should be established. The required compressive strength affects the amount of cement to be used as well as the maximum water allowed. The required surface finish frequently will control the ratio of coarse to fine aggregate. The extent of exposure to severe weather or other harsh environments will affect the durability requirements of the concrete mix design.

Concrete mixes are usually divided into two groups, namely, face and backup. Face mixes are usually composed of special decorative aggregates, and are frequently made with white or buff cement and/or pigments, where exposed aggregate surface finishes are desired. Backup mixes are usually composed of more economical local aggregates and gray cement and are used to reduce material costs in large units employing face mixes. Backup concrete mixes are also used where exposed aggregate or other special finishes are not required, and where the size and distribution of aggregate are not critical. In precast concrete units of complicated shapes and deep narrow sections, the face mix may be used throughout the member if procedures for separating the face and backup mixes become too cumbersome.

C4.1.1 Qualification of New Concrete Mixes

Accepted methods of mix proportioning are given in detail in the following publications:

1. Portland Cement Association:

   a. *Design and Control of Concrete Mixtures*.

2. American Concrete Institute:

   a. *Standard Practice for Selecting Proportions for Normal and Heavyweight and Mass Concrete (ACI 211.1)*.

   b. *Standard Practice for Selecting Proportions for Structural Lightweight Concrete (ACI 211.2)*.

   c. *Standard Practice for Selecting Proportions for No-Slump Concrete (ACI 211.3)*.

   d. *Specifications for Structural Concrete for Buildings (ACI 301)*.

Standard

Concrete mixes shall be proportioned and/or evaluated for each individual project with respect to strength, absorption, volume change, and resistance to freezing and thawing, where such environments exist, as well as desired surface finish (color and texture). The mix shall have adequate workability for proper placement and consolidation.

4.1.2 Specified Concrete Strength

Concrete strengths shall be determined on the basis of test specimens either at time of stripping or at a specified age, usually 28 days, although other ages may be specified. A minimum acceptable strength at time of stripping shall be established by the precast engineer and shall be stated on the drawings. When members are prestressed, the concrete shall have a specified compressive strength suitable for transfer of prestress at time of stripping and 28 day strength as required by the specifications, unless otherwise specified by the Engineer.

4.1.3 Statistical Concrete Strength Considerations

For commonly used concrete mixes such as backup mixes, or for face mixes where the size of the project warrants, a plant shall maintain up-to-date documentation of the compressive strength variability. Based on this information, a design strength shall be chosen for the concrete which will comply with the statistical interpretation of the strength requirements given in ACI 318.

Commentary

An architectural precast concrete mix has demanding criteria since its appearance will be a governing item, but strength and durability will also be required. Mix design factors which influence appearance are selection and proportioning of fine and coarse aggregate, color of cement and use of pigments. Often several aggregates must be blended and the properties of the mixture considered in proportioning the mix. Typically a larger portion of the mix will be coarse aggregate which affects consistency, finishability and strength. Careful selection of aggregate sizes, colors and mix proportions is needed to provide the desired emphasis upon the matrix or the exposed surface texture. These considerations vary significantly from those for proportioning for strength and durability alone and the use of samples becomes a necessary step in the mix design process.

C4.1.2 Specified Concrete Strength

A minimum design strength for concrete should be determined by the architect/engineer, based upon in-service requirements. Consideration for production and erection are the responsibility of the precaster. The mix is generally proportioned for appearance, and strength becomes a secondary consideration. Except for load-bearing units, stresses on units are usually higher during fabrication and erection than those anticipated in the structural design for in-service conditions. Production requirements for early stripping of units or early stress transfer and subsequent rapid reuse of forms demand high levels of early compressive strength. The minimum transportation and erection strength levels will depend on shape of the unit, handling, shipping and erection techniques and on schedule, which will normally result in 28-day strengths higher than the specified minimum.

In cases where the typical 28 day strength of 5000 psi (34.5 MPa) is not structurally necessary, or may be difficult to attain due to special cements or aggregates, sufficient durability and weathering qualities may often be obtained by controlling proper air entrainment and absorption limits at a strength level as low as 4000 psi (27.6 MPa).

C4.1.3 Statistical Concrete Strength Consideration

Concrete strength test evaluation should follow methods outlined in ACI 214.

Strength tests failing to meet these criteria may occur occasionally (probably about once in 100 tests), even though strength level and uniformity are satisfactory. Allowance should be made for such statistically normal deviations in deciding whether or not the strength level being produced is adequate.

Mix designs and concrete proportions may be selected on the basis of established records for the concrete production facility. The better the control, as measured by the coefficient of variation or standard deviation, the more economical the selected mix may become. Under the best control, there still are many variables that can influence concrete strength, such as variations of ingredients, variations in batching, and testing. Concrete for which all specimens can be expected to show strengths above the specified minimum strength is generally impractical, and evaluation of strength tests should recognize this fact.

Standard

The strength level of the concrete shall be considered satisfactory if the average of each set of any three consecutive strength tests equals or exceeds the specified strength and no individual test falls below the specified strength by more than 500 psi (3.5 MPa).

4.1.4 Proportioning to Ensure Durability of Concrete

Concrete strength and durability shall be achieved through proper consideration in the mix design of air, water and cement contents, and workability. Low water-cement ratios shall be used to provide specified strength, durability and low absorption. Drying shrinkage characteristics shall be controlled by aggregate size, gradation, mineralogy, aggregate-cement ratio, cement factor, water-cement ratio, additives, and admixtures. All of these above items shall be considered.

All combinations of cement and aggregates shall be those for which chemical and physical properties or long experience have shown conclusively that the various components are compatible and will not result in serious volume changes, cracking or deterioration of the concrete as it ages. In particular, the use of high alkali cement together with alkali reactive aggregates is prohibited. Also to be avoided are aggregates subject to "pop-outs", rusting, staining, or other surface deterioration. These precau-

Commentary

Concrete for background tests to determine standard deviation is considered to have been "similar" to that required if it was made with the same general types of ingredients under no more restrictive conditions of control over material quality and production methods than will exist on the proposed work, and if its specified strength did not deviate more than 1000 psi (6.9 MPa) from the required strength. A change in the type of cement or a major increase in the required strength level may increase the standard deviation.

Adequate statistical records are based on at least 30 consecutive strength tests obtained within the past year representing similar materials and conditions to those expected. The 30 consecutive strength tests may represent either a group of 30 consecutive batches of the same class of concrete or the statistical average for two groups totaling 30 or more batches.

Average strengths, used as the basis for selecting proportions, should exceed the specified strength by at least:

> 400 psi (2.76 MPa) if the standard deviation is less than 300 psi (2.07 MPa).
>
> 550 psi (3.79 MPa) if the standard deviation is 300 (2.07 MPa) to 400 psi (2.76 MPa).
>
> 700 psi (4.82 MPa) if the standard deviation is 400 (2.76 MPa) to 500 psi (3.45 MPa).
>
> 900 psi (6.20 MPa) if the standard deviation is 500 (3.45 MPa) to 600 psi (4.14 MPa).

If the standard deviation exceeds 600 psi (4.14 MPa) or if a suitable record of strength test performance is not available, proportions should be selected to produce an average strength at least 1200 psi (8.27 MPa) greater than the specified strength

C4.1.4 Proportioning To Ensure Durability of Concrete

Selection of fine and coarse aggregate affects both density and appearance. The high quality of the aggregate does not necessarily by itself assure good mix design, but it does relate to quality of concrete as evidenced by its durability. If the aggregate material is not durable, deterioration can be expected in areas where the concrete is in contact with corrosive atmospheres, salts, excessive moisture, and/or severe temperature changes.

Achieving low absorption rates for the surface of the concrete demands a high density of the concrete surface.

Standard

tions are particularly necessary in areas subject to freezing and thawing or in locations where salt or sulfate exposures are expected.

4.2 Special Considerations for Air Entrainment

Units subject to freezing and thawing shall be air entrained. For gap-graded facing mixes, where a given percentage of air cannot be reliably measured, the dosage of air-entraining agent shall produce an 8 to 10 percent air content when tested in accordance with ASTM C 185 but using only the mortar [material passing the No. 4 (4.75 mm) sieve] portion of the mix or provide 19% ±3% of entrained air in the paste when tested according to ASTM C 185. Once established for the mixture, the corresponding entrained air content of the total concrete mixture may be determined, and that value shall be used in production control.

When a specific level of air content is to be maintained in concrete units exposed to freeze-thaw, deicer and wet-dry conditions, air content at the point of delivery shall conform to the requirements of Table 4.2.1. For specified compressive strength greater than 5000 psi (34.5 MPa), reduction of air content indicated in Table 4.2.1 by 1.0 percent shall be permitted.

Table 4.2.1. Total air content for various sizes of coarse aggregate for normal weight concrete.

Nominal maximum size of aggregate, in. (mm)	Total air content, percent, by volume[1]	
	Severe exposure	Moderate exposure
Less than 3/8 (9)	9	7
3/8 (9)	7 1/2	6
1/2 (13)	7	5-1/2
3/4 (19)	6	5
1 (25)	6	5
1-1/2 (38)	5-1/2	4-1/2

1. Air content tolerance is ±1 1/2 percent.

The properties of the concrete-making materials, the proportioning of the concrete mixture, and all aspects of mixing, handling, and placing shall be

Commentary

C4.2 Special Considerations for Air Entrainment

Air-entrained concrete should be able to withstand the effects of freezing as soon as it attains a compressive strength of about 500 psi (3.45 MPa) provided that there is no external source of moisture. Because of the gradation characteristics of most gap-graded facing mix concrete or for mixes with high cement contents and low slumps, a given percentage of air cannot be reliably measured for many mixes.

Air is entrained in the mortar fraction of the concrete; in properly proportioned mixes, the mortar content decreases as maximum aggregate size increases, thus decreasing the required concrete air content for both workability and durability.

Typical plant control practice involves only the measurement of air volume in freshly mixed concrete. Although measurement of air volume alone does not permit full evaluation of the important characteristics of the air-void system, air-entrainment is generally considered effective for freeze-thaw resistance when the volume of air in the mortar fraction of the concrete (material passing the No. 4 sieve) is about 9 ±1%.

Standard

maintained as constant as possible in order that the air content will be uniform and within the range specified for the work.

4.3 Compatibility of Face and Backup Mixes

In order to assure compatibility of face and backup mixes the following shall be investigated for their consequences on the unit design: (1) relative shrinkage characteristics; (2) relative thermal coefficients of expansion; and (3) relative modulus of elasticity.

Special attention shall be given to mix compatibility when normal weight face mixes are combined with lightweight backup mixes.

4.4 Proportioning for Appearance of Concrete Surface

The face mix for architectural concrete units shall be designed to produce the desired appearance taking into account the technique for obtaining the surface finish.

4.5 Mix Proportioning for Concrete Made with Structural Lightweight Aggregate

Lightweight concrete proportions shall be selected to meet the specified limit on maximum air-dry unit weight as measured in accordance with ASTM C 567.

Commentary

C4.3 Compatibility of Face and Backup Mixes

Face and backup mixes should be proportioned to minimize variations in shrinkage, thermal coefficient of expansion, and modulus of elasticity such that finished products meet the tolerance requirements for bowing and warping. These mixes should have similar water-cement and cement-aggregate ratios.

The combination of a normal weight face mix and a backup with lightweight aggregates may increase the possibility of bowing or warping. Before using such a combination, sample units produced, cured and stored under anticipated production conditions, should be made to verify satisfactory performance.

C4.4 Proportioning for Appearance of Concrete Surface

The type of finish to be achieved is the key factor to consider when making the decision regarding the quantity of coarse aggregate desired in the architectural concrete face mix. The material requirements will vary with the depth and amount of surface removed. The shallower the reveal, the more visual influence the aggregate fines and cement will have, and the deeper the reveal the more visual influence the coarse aggregate will have.

Proportioning of fine and coarse aggregate is of importance to workability of the mix and, thus, to final appearance.

In general, where appearance and color uniformity are of prime importance, mixes may have considerably higher cement content than is normally required to achieve a specified strength. If the cement is the main contributor to the color of concrete, the color will become more intense or darker with increased cement content or decreased water-cement ratio.

C4.5 Mix Proportioning for Concrete Made with Structural Lightweight Aggregate

Mix proportioning methods for structural lightweight aggregate concrete (ACI 211.2) generally differ somewhat from those for normal weight concrete. The principal properties that require modification of proportioning and control procedures are the greater total water absorption and rate of absorption of lightweight aggregates, plus their low weight. The absorption of water by the aggregate has little effect on compressive strength, provided that enough water is supplied to saturate the aggregate. The moisture content of the aggregate must be known, and adjustments must be made from batch to batch to provide constant cement and air contents, similar slumps, and a constant volume of aggregates.

Standard

4.6 Proportioning for Concrete Workability

The slump and workability of a mix shall be suitable for the conditions of each individual job, i.e., to permit the concrete to be worked readily into the forms and around reinforcement under the conditions of placement to be used, without excessive segregation or bleeding. Workability considerations shall include the shape of the unit, height of casting, the amount and complexity of reinforcement, and the method of consolidation.

In color sensitive units, the slump shall be as low as possible and constant from batch to batch in order to provide uniformity of color in the end product. Slump tolerances of ±1 in. (±25 mm) prior to the addition of HRWR shall be maintained for mix consistency and color control.

When superplasticized concrete falls below the specified slump due to a delay, it shall be retempered with superplasticizing admixtures only rather than additional water.

4.7 Water-Cement Ratio

4.7.1 General

The water-cement ratio shall not exceed 0.45 by weight with an allowable variation during production of ±0.02. In all mix designs for precast concrete production the importance of a low water-cement ratio shall be recognized, but this ratio shall also be evaluated in relation to the workability required for satisfactory placing and consoli-

Commentary

C4.6 Proportioning for Concrete Workability

It is difficult and unnecessary to establish limits for slump for typical precast concrete production. Very stiff mixes require more labor to place and special vibration techniques.

Required workability is related to the shape of the precast concrete unit and the method of consolidation. Methods of consolidation using mechanical equipment require less fines for workability. The lower the percentages of fines, the lower the specific surface area of the aggregates and the lower the water and cement requirements of the mix, which leads to a higher quality potential of the concrete. However, very low fines content mixes may bleed excessively and require a high water content for workability.

Workability under the influence of a properly selected vibrator is the important consideration and not slump. Workability of freshly mixed concrete is the property that determines the ease and homogeneity with which it can be mixed, placed, compacted, and finished.

Slump is a measure of concrete consistency. However, it is not, by itself, a measure of workability. Other considerations such as cohesiveness, harshness, segregation, bleeding, ease of consolidation, and finish ability are also important, and these properties are not entirely measured by slump.

The consistency of concrete, as measured by the slump test, is an indicator of the relative water content of the same concrete mixture. Additional water increases the water-cement ratio and has the undesirable effect of reducing the cohesion within the mixture and increasing the potential for segregation and excessive bleeding.

Excessive water will affect strength, shrinkage, density and absorption and uniformity of color.

Concrete slumps in color sensitive architectural panels should not exceed 3 in. (75 mm) in most instances, prior to the addition of high range water reducer, exceptions to this are gap-graded mixes which should generally not have slumps greater than 2 in. (50 mm)

C4.7 Water-Cement Ratio

C4.7.1 General

With given materials, the optimum mixture proportions use the least amount of total water per unit volume of concrete to obtain the required slump and workability. With a fixed water-cement ratio, material costs are reduced by using mixtures having the least paste. The cement in the paste is typically the most costly ingredient of the concrete; therefore, using more paste than required adds unnecessarily to the cost of the concrete. However, when using extremely high-priced aggregates, the

Standard

dation techniques for the actual mix application. Water shall be limited to the minimum needed for proper placing and consolidation by means of vibration. Low water-cement ratios shall be accompanied by controls on total water content.

4.7.2 Relationship of Water-Cement Ratio to Strength, Durability and Shrinkage

Since the water-cement ratio is one of the fundamental keys governing the strength and durability of the concrete the proportioning of the concrete mix design process shall minimize the water-cement ratio to the maximum extent possible consistent with acceptable workability of the concrete mix in the intended application.

4.7.3 Relationship of Water-Cement Ratio to Workability

The water-cement ratio shall not be increased for reasons of improving workability. Use of suitable workability improving admixtures shall be employed if additional workability is needed in concrete mixes which are already proportioned at the maximum allowable water-cement ratio.

4.8 Effects of Admixtures

All types of admixtures used shall be materials of standard manufacture having well established records of tests to confirm their properties and their short term and long term effects on the properties of both fresh and cured concrete. The manufacturers' recommendations shall be followed in their use.

Whenever more than one admixture is used in a concrete mix, it shall be verified prior to production that each material performs as required without adversely affecting the performance of the other.

Commentary

cost of paste should be balanced against the aggregate cost.

Most chemical admixtures of the water-reducing type are water solutions. The water they contain becomes a part of the mixing water in the concrete and should be considered in the calculation of water-cement ratio. The proportional volume of the solids included in the admixture is so small in relation to the size of the batch that it can be neglected in the mix design calculations.

When using a silica fume slurry, the water portion of the slurry must be taken into account when determining mix proportions. This entails reducing the amount of batch water to correct for the water in the slurry.

C4.7.2 Relationship of Water-Cement Ratio to Strength, Durability and Shrinkage

A reasonable balance should be established between a maximum cement content for stripping and service strength requirements and a minimum cement content to diminish its negative qualities, such as shrinkage and a matrix hardness lower than that of the aggregates.

Minimizing the paste is desirable because water in the paste is the primary cause of shrinkage as the concrete hardens and dries. The more water (i.e., the more paste), the greater the drying shrinkage. Also, cement produces heat as it hydrates. Therefore, high cement contents may produce an undesirable temperature rise during curing and crack-producing temperature differentials.

C4.7.3 Relationship of Water-Cement Ratio to Workability

The water content of concrete is influenced by a number of factors: aggregate size and shape, slump, water-cement ratio, air content, cement content, admixtures, and environmental conditions. Increased air content and aggregate size, reduction in water-cement ratio and slump, rounded aggregates, and the use of water-reducing admixtures reduce water demand. On the other hand, increased temperatures, increased cement contents, increased slump, increased water-cement ratio, aggregate angularity, and a decreased proportion of coarse aggregate to fine aggregate increase water demand.

C4.8 Effects of Admixtures

The development of admixtures to modify and improve the properties of fresh and cured concrete is one of the most rapidly changing areas of concrete technology. Some concrete admixtures have been in successful service for many years and can be considered well proven performers with no adverse side effects.

Standard

Regardless of the form of the admixture, provision shall be made for controlling the quantity and uniform introduction with other concrete components to ensure it's well distributed incorporation into the mix.

4.9 Storage and Handling of Concrete Materials

4.9.1 General

Concrete batching plants and their operation shall be in conformance with ASTM C94. Concrete batch plants shall be capable of producing concrete of the quality required for architectural precast concrete members; and they shall be properly equipped, maintained and operated. Batching and mixing facilities shall have provisions for producing concrete during the expected range of hot and cold ambient temperatures normally encountered. There shall be an adequate water supply, with pressures sufficiently constant or regulated to prevent interference with accuracy of measurement.

Concrete supplied to a plant by an outside batch plant shall meet the same requirements of on site batch plant facilities. Evidence of conformance shall be certification of the outside plant by NRMCA (National Ready Mixed Concrete Association).

4.9.2 Storage and Handling of Aggregates

Each aggregate shall be handled and stored by methods that will minimize variability in grading and moisture content upon arrival at batch weighing equipment. Rehandling of aggregates shall be minimized to avoid the aggregate segregation which can occur as result of each handling operation.

Wet or moist aggregates shall be stockpiled in sufficient time before use so that they can drain to uniform moisture contents. Required storage times, depending primarily on the grading and particle shapes of the aggregate, shall be verified by moisture tests or measurements.

Stockpiles. When aggregates are to be stockpiled, the use of aggregate bins is preferred; but failing this, it is imperative that a hard, clean and well-drained base shall be provided for each aggregate stockpile. If contamination from underlying material cannot otherwise be avoided, the area shall be planked or paved. Stockpiles shall be built up in horizontal or gently sloping layers. Conical stockpiles or any unloading procedure involving the dumping of aggregate down sloping sides of piles

Commentary

C4.9 Storage and Handling of Concrete Materials

C4.9.2 Storage and Handling of Aggregates

Procedures for handling and storage of aggregates are outlined in further detail in ACI 221 and ACI 304.

Unless uniformity of aggregates as batched is assured, production of uniform concrete is unlikely. It is essential that aggregates be kept clean and that contamination of one aggregate by another is prevented.

Stockpiles. Stockpiles of coarse aggregate inevitably tend to accumulate an excess of fines near their bases. Enough of this fine material should be periodically removed and discarded so the coarse aggregate as delivered to the batching bins is within specified gradation limits.

Standard

shall be prohibited. A front-end loader or reclaimer shall remove slices from the edges of the pile so that every slice will contain a portion of each horizontal layer. Trucks, bulldozers, or other vehicular or track equipment shall not be operated on the stockpiles because in addition to breaking the aggregate, dirt is frequently tracked onto the piles. Intermixing of different materials shall be prevented by suitable walls or ample distance between piles.

Fine aggregate shall be handled in a damp state to minimize the separating of dry fine aggregate by wind action. Stockpiles shall not be contaminated by spillage from swinging aggregate filled buckets, conveyor belts, or clams over the various stockpiled aggregate sizes.

Bins. When bins are used for storing aggregates, separate compartments shall be provided for fine aggregates and for each required size of coarse aggregate. Each compartment shall be capable of receiving and storing material without cross contamination. Bins shall be filled by material falling vertically, directly over the outlet. Each compartment shall be designed to discharge freely and independently into the weigh hopper. Bins shall be kept as full as practicable at all times.

Storage bins shall have the smallest practicable equal horizontal dimensions. To avoid accumulation of fines in dead storage areas, bottoms of circular bins shall slope at angles not less than 50 degrees from the horizontal toward center outlets. The bottom slope of rectangular bins shall slope at angles not less than 55 degrees from the horizontal.

Bags. When bagged aggregates are used, the individual sizes shall be stored on pallets in a well drained, reasonably dry area. Bagged fine aggregate shall be stored under dry conditions. Bagged aggregate shall be weighed prior to usage.

4.9.3 Storage and Handling of Cement

Any cement that develops hard lumps (due to partial hydration or dampness) which cannot be reduced by light finger pressure shall not be used, unless tested for strength or loss-on-ignition

Bulk Cement. Bulk cement shall be stored in weather-tight bins or silos which exclude moisture and contaminants. Storage silos shall be drawn down frequently, at least once per month, to prevent cement caking. Each brand, type and color of

Commentary

Bins. Aggregate bins in cold climates may have to be appropriately heated in winter.

Chuting the aggregate into a bin at an angle and against the bin sides will cause it to segregate. Baffle plates or dividers will help minimize segregation. Round bins are preferred.

By keeping bins full, breakage and changes in grading will be minimized as the materials are drawn down.

Bags. If stored outside, aggregates bagged in burlap should be protected with a weather protective cover to prevent deterioration of the bags from moisture in wet regions. If aggregates are to be stored in polypropylene bags, the bags should be protected from sunlight to prevent deterioration of the bags. Long periods of storage may require rebagging or other means to prevent breaking of bags when handled. Storage should be such that there is no mixing of sizes if some bags break and the aggregate is recovered.

C4.9.3 Storage and Handling of Cement

Cement can be supplied in bags or in bulk. Portland cement has great affinity for water and if left exposed to the atmosphere will gradually absorb water from the air and thereby become "set" into small lumps. If kept dry it will retain its quality indefinitely.

Bulk Cement. Most of the contamination of cements occurs during shipping and handling. It is generally caused by failure to clean out trucks and rail cars in which cement is to be shipped. Changes in color or texture or the presence of coarse particles may be evidence of a problem.

Standard

cement shall be stored in a separate bin.

Silos and Bins. Compartments shall be designed to discharge freely and independently into the weighing hopper. The interior of a cement silo shall be smooth with a minimum bottom slope of 50 deg. from the horizontal for a circular silo and 55 deg. for a rectangular silo. Silos not of circular construction shall be equipped with features to loosen cement which has settled tightly in the silos.

Each bin compartment from which cement is batched shall include a separate gate, screw conveyor, air slide, rotary feeder, or other conveyance which effectively combines characteristics of constant flow with a precise cut off feature to obtain accurate automatic batching and weighing of cement.

Procedures shall be in place to avoid cement being transferred to the wrong cement silo, either by faulty procedures or equipment.

Bagged Cement. Cement in bags shall be stacked on pallets or similar platforms to avoid contact with moisture and to permit proper air circulation. Bags shall be stored clear of wall areas where condensation may occur. Bags to be stored for long periods shall be covered with waterproof coverings. Bags shall be stacked so that the first in are the first out.

4.9.4 Storage and Handling of Admixtures and Pigments

All manufacturers of admixtures and pigments have specific recommendations for storage, handling, batching and use of their particular materials. These instructions shall be followed at all times. Adequate storage facilities shall be provided to ensure uninterrupted supplies of admixtures during batching operations. Liquid admixtures shall be stored separately in weather-tight containers or tanks, with each one clearly labeled with type, brand and manufacturer.

Provision shall be made for proper venting so that foreign materials cannot enter storage tanks or drums through openings and tanks or drums do not become air bound, restricting the admixture flow. Facilities for straining, flushing, draining, and cleaning the storage tanks shall be provided. Fill nozzles and other tank openings shall be capped when not in use to avoid contamination.

In addition to mechanical or electromechanical dispensing systems used for measuring and charging of the admixtures to the concrete batch, a calibrated holding tank shall be part of each system so that plant operators can visually verify that the

Commentary

Bagged Cement. Portland cement should be kept sealed in its original bags and well protected from water or humidity until use. Mixes should be sized to use full bags of cement whenever possible. If it is necessary to store partially used bags of cement, they should be folded closed and then enclosed in a polyethylene bag. Old cement may reduce the strength of the concrete.

C4.9.4 Storage and Handling of Admixtures and Pigments

The requirements for storage of powdered admixtures are the same as those for storage of cement.

Admixture manufacturers usually furnish either complete storage and dispensing systems or at least information regarding the degree of agitation or recirculation required for their admixtures. Timing devices commonly are used to control recirculation of the contents of storage tanks to avoid settlement or, with some products, polymerization.

High volume liquid admixtures, such as non-chloride accelerators or silica fume slurries may not use a calibrated holding tank. They may be metered directly into the mixer.

Standard

proper amounts of each admixture are batched.

All admixture dispensers shall provide for diversion of the measured dosages for verification of the batch quantity. Batching accuracy shall be checked at least every 90 days. Calibrated sight tubes shall be vented so that they do not become air bound and restrict flow. Piping for liquid admixtures shall be free from leaks and properly fitted with valves to prevent backflow or siphoning and to ensure that measured amounts are completely discharged.

Every volumetric admixture dispenser shall be provided with visual indication or interlock cutoff when its liquid admixture supply is depleted. Dispenser control panels shall be equipped with timer-relay devices to ensure that all admixtures have been discharged from the conveying hoses or pipes.

Tanks, lines, and dispensing equipment for liquid admixtures shall be protected and configured to prevent freezing, contamination, dilution, evaporation and shall have a means for preventing settlement or separation of the admixtures. To prevent freezing, storage tanks and their contents shall be either heated or placed in heated environments. The manufacturers' instructions regarding the effects of heat or freezing on the admixtures shall be observed.

Separate dispensers shall be used for each admixture, although multiple use of dispensing controls is permitted. Compatible admixtures may be stored in the same holding or checking reservoir after batching and prior to introduction into the mixer. If the same dispensing equipment is used for non-compatible admixtures, the dispenser shall be flushed at the end of each cycle.

Storage areas for pigments shall be adequately clean and dry. Pigments shall remain in sealed containers until used, and all opened packages shall be protected from contamination.

Silica fume shall be packaged in bags or in drums (slurry). Silica fume slurries in drums will stiffen or gel during storage without appreciable settling and shall be remixed before being used. Bulk storage tanks shall be equipped with a mechanical agitation device or with a recirculation system that is designed to appropriately remix all of the silica fume. Metakaolin shall be packaged in powder form in bags or bulk.

Commentary

Some admixtures become quite viscous at lower temperatures which might cause difficulty in their use in cold weather. Freezing can cause ingredients of some liquid admixtures to separate and, therefore, affect concrete quality control.

Pigments are packaged in bags (dry powder pigments) and drums (liquid pigments).

Standard

4.10 Batching Equipment Tolerances

Batching equipment shall be maintained and operated in accordance with ASTM C94.

The quantities of ingredients used for each batch shall be recorded separately for each batch.

When measuring by bulk volume, batching shall be in accordance ASTM C 685 with the weight tolerance waived.

If graphical recorders are used, they shall register scale readings within ± 2 percent of total scale capacity, while digital recorders shall reproduce the scale reading within ± 0.1 percent of scale capacity.

For ingredients batched by weight, the accuracy tolerances required of the batching equipment shall be applicable for batch quantities between 10 and 100 percent of scale capacity.

For water or admixtures batched by volume, the required accuracy tolerances shall be applicable for all batch sizes from minimum to maximum, as is determined by the associated cement or aggregate batcher rating.

Operation and maintenance of batching equipment shall be such that the concrete ingredients are consistently measured within the following tolerances:

For **Individual Batchers,** the following tolerances shall apply based on the required scale reading:

Cement and other Cementitious Materials:

±1 percent of the required weight of material being weighed, or ± 0.3 percent of scale capacity, whichever is greater.

Aggregates:

± 2 percent of the required weight of material being weighed, or ± 0.3 percent of scale capacity, whichever is greater.

Water:

± 1 percent of the required weight of material being weighed, or ± 0.3 percent of scale capacity, or ± 10 lbs, whichever is greater.

Admixtures:

± 3 percent of the required weight of material being weighed, or ± 0.3 percent of scale capacity, or ± the minimum dosage rate for one 94 lb (43 kg) bag of cement, whichever is greater.

Pigments in powder form are used in extremely small dosages and shall be batched by hand from

Commentary

C4.10 Batching Equipment Tolerances

The range of tolerance control, weight control limitations of batching equipment and batching controls are covered in the Concrete Plant Mixer Standards of the Concrete Plant Manufacturers Bureau.

Standard

pre-measured containers packaged in amounts sufficient for proper dosages per unit volume of concrete. Pigments shall be weighed to the nearest ± 1% of the required weight.

For **Cumulative Batchers** without a tare compensated control, the following tolerances shall apply to the required cumulative scale reading:

Cement and other Cementitious Materials or Aggregates:

± 1 percent of the required cumulative weight of material being weighed, or ± 0.3 percent of scale capacity, whichever is greater.

Admixtures:

± 3 percent of the required cumulative weight of material being weighed, or ± 0.3 percent of scale capacity, or ± the minimum dosage rate per bag of cement, whichever is greater.

For **Volumetric Batching Equipment** the following tolerances shall apply to the required volume of material being batched:

Water:

± 1 percent of the required weight of material being batched, or ± 1 gallon, whichever is greater.

Admixtures:

± 3 percent of the required volume of material being batched but not less than ± 1 ounce (30 gm) or ± the minimum recommended dosage rate per 94 lb. (43 kg) bag of cement, whichever is greater.

Aggregates

When measuring lightweight aggregate by bulk volume, batching shall be in accordance with ASTM C685 with the weight tolerance waived.

4.11 Scale Requirements

Each scale in the plant shall consist of a suitable system of levers or load cells that weigh consistently within specified tolerances, with loads indicated either by means of a beam with balance indicator, a full-reading dial, or a digital read-out or display.

For all types of batching systems, the batch opera-

Commentary

Water:

The mechanisms in most commercial water meters cannot respond to quantities less than 1-gallon increments.

Admixtures:

Dispensers for liquid admixtures may measure either by volume or weight. Powdered admixtures should be measured by weight. Generally better results are obtained from admixtures in liquid form. Modern liquid admixture batching equipment incorporating effective controls and interlocks have virtually eliminated the requirement for weighing admixtures in the powdered state.

When high-volume usage of pigments is contemplated, the supplier of the pigment may supply a suitable bulk dispensing system.

Aggregates:

In some instances the accurate control of concrete with lightweight aggregate is more feasible measuring by bulk volume rather than by weight.

Standard

tor shall be able to read the load-indicating devices from the operator's normal station. Where controls are remotely located with respect to the batching equipment, monitors or scale-follower devices shall repeat the indication of the master scale within ±0.2 percent of scale capacity.

Separate scales shall be provided for weighting cement and other cementitious materials. Gates, valves, controlled screw feeders, air slides or other effective devices that will permit a precise cutoff for cement shall be used to charge the cement weighing equipment.

The reading face capacity or the sum of weigh-beam capacities of a scale on a cement batcher shall be not less than 660 pounds per cubic yard (390 kg per m^3) of rated batcher capacity.

Fine and coarse aggregates shall be weighed on separate scales or on a single scale which will first weigh one aggregate, then the cumulative total of aggregates. The reading face capacity or the sum of weigh-beam capacities of a scale on an aggregate batcher shall be not less than 3,300 pounds per cubic yard (5236 kg per m^3) of rated batcher capacity.

All scales shall be maintained so they are capable of accuracy within 1 percent of loads weighed under operating conditions or within ± 0.20 percent of scale capacity throughout the range of use. For direct digital read-out, the tolerance shall be increased to ± 0.25 percent, to allow for the fact that digital readings may be limited to whole-number values which cannot reproduce weight indications closer than ± 0.05 percent of capacity. All exposed fulcrums, clevises and similar working parts shall be kept clean. Beam type scales shall be checked to zero load with the bins empty each time the mix is changed and at least once during each day's operation. Scales shall register loads at all stages of the weighing operations from zero to full capacity.

For calibration of scales, standard test weights aggregating at least 500 lb. (227 kg) (each accurate within ± 0.01 percent of indicated values) shall be used. Calibration of scales shall be performed at intervals not greater than 6 months, and whenever there is reason to question their accuracy. Scale calibration certificates and charts shall be displayed prominently at the batch control location.

4.12 Requirements for Water Measuring Equipment

The reading face capacity or the sum of weigh-beam capacities of a scale on a water batcher shall

Commentary

C4.12 Requirements for Water Measuring Equipment

Water meters should conform to the Standards of the American Water Works Association.

Standard

be not less than 320 pounds (145.5 kg) or 38 gallons per cubic yard (188 liters per cubic meter) of rated batcher capacities.

If added water is to be measured by volume, the water measuring device shall be arranged so the measurements will not be affected by variable pressures in the water supply line.

4.13 Requirements for Batchers and Mixing Plants

4.13.1 General

It is not the purpose of this Article to specify any particular type of concrete plant. The batch plant which supplies concrete to the precast concrete fabricating activity shall be so equipped that batching, mixing, and transporting equipment provide sufficient quantities of concrete to maintain the casting schedule. The concrete mix shall be properly proportioned and mixed to the desired uniformity and consistency, and be capable of developing the required strength.

Batchers for weighing cement, aggregates, water and admixtures (if measured by weight) shall consist of suitable containers freely suspended from scales, equipped with necessary charging and discharging mechanisms. Batchers shall be capable of receiving rated load without contact of the weighed material with the charging mechanisms.

Charging or dispensing devices shall be capable of controlling the rate of flow and stopping the flow of material within the weighing tolerances. Charging and discharging devices shall not permit loss of materials when closed. Systems shall contain interlocks which prevent batcher charging and discharging from occurring simultaneously and provide that, in the event of electrical or mechanical malfunctions, materials cannot be over batched. Provision shall be made for removal of material

Commentary

Measuring tanks for water should be equipped with outside taps and valves to provide for checking meter calibration not less frequently than every 90 days. Volumetric tank water batchers should be equipped with a valve to remove overloads.

In the case of truck mixers used to supply the plant, if wash water is permitted to be used as a portion of the mixing water (back up mixes only - don't use wash water for face mixes) for the succeeding batch, it shall be measured accurately in a separate tank and taken into account in determining the amount of additional water required.

C4.13 Requirements for Batchers and Mixing Plants

C4.13.1 General

Concrete batching plants include simple manual equipment in which the operator sets batch weights and discharges materials manually; semi-automatic plants in which batch weights are set manually and materials are discharged automatically; or fully automated electronically controlled plants in which mixes are controlled by means of selectors or punch cards or computer memory.

Any of the methods described are acceptable as long as concrete is of consistent quality and operations conform to ASTM C94.

When batch plants of outside suppliers are used they should be reviewed regularly by plant staff to confirm compliance with National Ready Mixed Concrete Association Plant Certification requirements and other applicable requirements of this division.

Standard

overloads.

The batchers shall be equipped with provisions to aid in the smooth and complete discharge of the batch. Vibrators or other appurtenances shall be installed in such a way as not to affect accuracy of weighing, and wind protection shall be sufficient to prevent interference with weighing accuracy.

Cementitious materials batchers shall be provided with a dust seal between charging mechanism and batcher, installed in such a manner that it will not affect weighing accuracy. The weigh hopper shall be vented to permit escape of air. The hopper shall be self-cleaning and fitted with mechanisms to ensure complete discharge of the batch.

4.13.2 Requirements for Concrete Mixers

Mixing equipment shall be of capacity and type to produce concrete of uniform consistency, and uniform distribution of materials as required by ASTM C94.

High intensity stationary mixers, of the vertical or horizontal shaft type capable of producing uniform low slump concrete, shall be used for architectural concrete face mixes. Backup mixes may be mixed by any of the methods listed in Article 4.17.2.

Mixers with a rated capacity of 1 cu. yd. (0.8 m^3) or larger shall conform to the requirements of the Standards of the Concrete Plant Mixer Standards, Mixer Manufacturers Division of the Concrete Plant Manufacturers Bureau . Truck mixers shall conform to the requirements of the Truck Mixer and Agitator Standards of the Truck Mixer Manufacturers Bureau. Truck mixers shall be equipped with counters by which the revolutions of the drum, blades or paddles may be verified.

4.13.3 Mixer Requirements

For each mixer and agitator, the plant shall have on file, the capacity of the drum or container in terms of the volume of mixed concrete.

A mixer shall not be used to mix quantities greater than those recommended by the mixer manufacturer. The rate of rotation of drum or blades for sta-

Commentary

C4.13.2 Requirements for Concrete Mixers

An important aspect of mixer performance is batch-to-batch uniformity of the concrete which is also largely affected by the uniformity of materials and their measurements as well as by the efficiency of the mixer. Visual observation of the concrete during mixing and discharge from the mixer is an important aid in maintaining a uniform mix, particularly uniform consistency. Some recording consistency meters, such as those operating from the amperage draw on the electric motor drives for revolving drum mixers, have also occasionally proved to be useful. However, the most positive control method for maintaining batch-to-batch uniformity is a regularly scheduled program of tests of the fresh concrete including unit weight, air content, slump, coarse aggregate content, and temperature.

For a description of the various mixer types, refer to the Concrete Plant Mixer Standards, Mixer Manufacturers Division of the Concrete Plant Manufacturers Bureau.

C4.13.3 Mixer Requirements

Low-slump concrete may require smaller loads than rated mixer capacity.

Standard

tionary or truck mixers shall be in accordance with manufacturer's recommendations.

4.13.4 Maintenance Requirements for Concrete Mixers

All mechanical aspects of mixers or agitators, such as water measuring and discharge apparatus, condition of blades, speed and rotation of the drum or blades, general mechanical condition of the units and cleanliness shall be checked daily. Mixers shall be examined daily for changes in condition due to accumulation of hardened concrete or mortar, or due to wear of blades. Accumulations of hardened concrete shall be removed.

4.14 Concrete Transportation Equipment

4.14.1 General

Properly designed bottom dump buckets or hoppers shall be used to allow placement of concrete at lowest practical slump consistent with consolidation by vibration. Discharge gates shall have a clear opening no less than one-third the maximum interior horizontal area or five times the maximum aggregate size being used. If dump buckets are used side slopes of transportation containers shall be steep, being no less than 60 degrees from the horizontal. Controls on the gates shall permit personnel to open and close them during any portion of the discharge cycle.

All conveying equipment shall be thoroughly cleaned at frequent intervals during a prolonged casting, between casting of different mixtures, or between stages of the same units such as between facing mix and backup concrete.

4.14.2 Requirements for Concrete Agitating Delivery Equipment

Agitators, truck mixers or truck agitators, shall be capable of maintaining the concrete in a thoroughly mixed and uniform mass and of discharging the concrete with a satisfactory degree of uniformity. All types of mixers and agitators shall be capable of discharging concrete at its specified slump.

4.15 Placing and Handling Equipment

Placing equipment shall be capable of handling concrete of designed proportions so it can be readily consolidated by vibration. Account shall be taken of any special degree of harshness or stiffness of the mixes used, of any unusual variations in shape and position of casting, and of obstruc-

Commentary

C4.15 Placing and Handling Equipment

Adequate attention to placing and handling equipment is necessary to allow efficient handling of relatively low-slump concrete, particularly gap-graded concrete, which will not readily flow down a chute, drop out of a bucket or hopper, or discharge through gates, although it is workable in place when properly vibrated.

Standard

tions resulting from any unusual items to be cast into the concrete. The assessment of any special method or procedure shall be based on the uniformity and the quality of the end product.

Limitations on concrete consistencies and proportions shall not be imposed by inadequate chutes, hoppers, buckets or gates.

Chutes shall be amply steep, metal or metal-lined, round bottomed, of large size, rigid, and protected from overflow. Discharge gates, bucket openings and hoppers shall be large enough to allow concrete of the lowest slump likely to be used to pass quickly and freely. Chutes shall have a slope not exceeding 1 vertical to 2 horizontal and not less than 1 vertical to 3 horizontal.

Chutes having a length that causes segregation [more than 20 ft (0.3 m)] or having a slope greater than 1 vertical to 2 horizontal may be used if concrete materials are recombined by a hopper or other means before distribution.

4.16 Batching and Mixing Operations

4.16.1 General

All concrete shall be batched and mixed so that it represents a uniform mixture, with a uniform appearance, and with all of the ingredients in the designed amounts evenly distributed throughout the mixture.

Once an appropriate mix design has been developed and confirmed by laboratory testing or by other means to assure the suitability of the mix for the intended purpose, the constituent materials shall be accurately, uniformly, and consistently combined to produce a uniform mixture with the desired properties.

To produce concrete of uniform quality, the ingredients shall be measured accurately for each batch.

Commentary

C4.16 Batching and Mixing Operations

C4.16.1 General

The aim of all batching and mixing procedures is to produce uniform concrete containing the required proportions of materials. To attain this, it is necessary to assure the following:

1. The equipment provided will accurately batch the required amounts of material and the amounts can be easily changed, when required.

2. The required proportions of materials are maintained from batch to batch.

3. All materials are introduced into the mixer in proper sequence.

4. All ingredients are thoroughly intermingled during mixing, and all aggregate particles are completely coated with cement paste.

5. The concrete, when discharged from the mixer, will be uniform and homogeneous within each batch and from batch to batch.

Batching is the process of weighing or volumetrically measuring and introducing into the mixer the ingredients for a batch

Standard

In addition to accurate measurement of materials correct operating procedures for the batching and mixing operation shall also be used. Care shall be taken to ensure that the weighed materials are properly sequenced and blended during charging of the mixers to maintain batch-to-batch uniformity.

Materials shall enter the mixer at a point near the center of the mixer.

Information on mix designs, batching and mixing procedures shall be recorded and maintained as required for other quality control records as outlined in Division 6.

Ready mixed concrete delivered to the plant via truck from off-site batching locations shall be documented with tickets showing batch quantities, admixtures, mix designation, design slump and time of batching.

4.16.2 Batching of Aggregates

Low variation of aggregate grading as batched is required for the production of uniform concrete. This is a necessary prerequisite, even when a high order of accuracy in measurement and superior mixing and placing are provided.

Under normal conditions all of the aggregates shall be batched to the mixer after the initial 5 to 10 percent of the mix water has been charged. Eighty percent of the water is added at this same time. When heated water is used in cold weather, the addition of the cement shall be delayed until most of the aggregate and water have intermingled in the drum.

Batch weight of normal weight aggregates shall be based on the required weight of saturated surface-dry aggregate corrected for the moisture conditions at the time of batching. The weight setting system shall provide adjustments of aggregate batch weights for aggregate moisture, and the moisture meter shall be able to detect changes of 1 percent in the moisture content of fine aggregate. The batch weight tolerance shall apply to the accuracy of measurement of the corrected weight.

If moisture meters are not used, the free moisture of the fine aggregate shall be determined at least daily, or any time a change in moisture content becomes apparent. Corrections based on the results of the tests shall be made. Moisture testing shall be per ASTM C70 and ASTM C566.

Refer to Article 4.17.7 for lightweight aggregate batching.

Commentary

of concrete. Consistency in all phases of batching and mixing is a key to having a quality finished product. Each mix should be batched in the same sequence and then mixed the same length of time. Water content, dispersion of color and cement, and slump are all controlled through uniform batching and mixing.

C4.16.2 Batching of Aggregates

When heated water is used in cold weather, the order of charging may require some modification to prevent possible rapid stiffening.

Cement can fluff up as much as 35 percent when aerated for bulk handling, and sand with intermediate amounts of surface moisture can bulk and occupy more space in a bin or stockpile than a very dry or very wet sand. It is for this reason that batching of cement and aggregates is done by weight for the most part.

Standard

4.16.3 Batching of Cement and Pigments

Cement and pigments shall be batched in a manner that assures uniform distribution in the mix. Cement and coloring pigments shall be charged with the fine and coarse aggregates but shall enter the stream after approximately 10 percent of the aggregate is in the mixer. Free fall of cement shall not be permitted. Cement shall flow from its hopper into the stream of aggregates through a suitable enclosure chute, usually a tube of canvas or rubber.

If batching by bag, the weight of full bags of cement shall be checked for conformation to stated weight once for every 10 bags used and to determine that tolerances are being met. Any fractional bags shall not be used unless they are weighed.

4.16.4 Batching of Water

When mixing normal weight concrete, a portion of the mixing water, between 5 and 10 percent, shall precede and a like quantity should follow discharge of other materials. The remainder of the water shall be introduced uniformly with other materials. Charging of water shall be complete within the first 25 percent of the mixing time. Where mixers of 1 cu yd (0.8 m^3) capacity or less are used, the aggregates may be placed into the mixer first and then the cement and water introduced at the same time.

4.16.5 Batching of Admixtures

Admixtures shall be charged into the mix as solutions and the liquid shall be considered part of the mixing water. Admixtures that cannot be added in solution may be either weighed or measured by volume as recommended by the manufacturer. A procedure for controlling the rate of admixture addition to the concrete batch and the timing in the batch sequence shall be established and followed. Admixture discharge shall ensure uniform distribution of the admixture throughout the concrete mixture during the charging cycle.

Admixtures shall be charged to the mixer at the same point in the mixing sequence batch after batch in accordance with the admixture manufacturer's recommendations. Regardless of whether in liquid, paste or powdered form, the introduction of admixtures shall generally be at a rate proportional to that of the other concrete components to assure uniform distribution into the mix. Liquid chemical

Commentary

C4.16.3 Batching of Cement and Pigments

Of the constituent elements of the concrete mixture the proportioning of the cement in the mixture is the most important determinant of the characteristics of the concrete. The provisions outlined here for batching and mixing the cement recognize the importance of the cement in the mixture and should be observed.

C4.16.4 Batching of Water

Next to the cement, the amount of water in the concrete mixture has the most significant effect on the properties of the concrete. Because of the adverse consequences of the addition of excess water to the mix at various stages of the manufacturing process, stringent controls on the handling, measurement, and introduction of water to the mix should be observed.

C4.16.5 Batching of Admixtures

Most admixtures are furnished in liquid form and often do not require dilution or continuous agitation to maintain their solution stability. For ease of handling and increased precision in batching, admixtures are preferable in liquid form.

If the admixture is supplied in the form of powder, flakes, or semisolids, a solution should be prepared prior to use, following the recommendations of the manufacturer. When this is done, mixing drums or storage tanks from which the admixture will be dispensed should be equipped with agitation or mixing equipment to keep solids in suspension.

Because of the fact that small quantities of admixtures, and combinations of admixtures, can create large changes in the properties of the fresh and cured concrete, the handling, storage, and measurement of admixtures, which have been given special attention in this document should be observed.

In some instances, changing the time at which the admixture is added during mixing can vary the degree of effectiveness of the admixture. This may cause the water demand to vary which may cause inconsistent finish and color. For some admixture-cement combinations, varying the time at which they are added during mixing may result in varying degrees of retardation or acceleration. Varying dosage rates in different batches may cause color variations.

Standard

admixtures shall not be added directly to the cement or to dry, absorptive aggregate. Liquid admixtures shall, in most cases, be charged with the last portion of mix water. The entire amount of non-retarding admixtures shall be added prior to the completion of the addition of the mixing water. Addition of retarding admixtures shall be completed within 1 min. after addition of water to the cement has been completed, or prior to the start of the last three-fourths of the mixing cycle, whichever occurs first.

If two or more admixtures are used in the concrete, they shall be added separately to avoid possible adverse interaction. This practice shall be followed unless tests indicate there will be no adverse effects or the manufacturer's advice permits intermixing of admixtures.

Tanks, conveying lines, and ancillary equipment shall be drained and flushed on a regular basis per manufacturer's recommendations. Calibration tubes shall have water fittings installed to allow the plant operator to water flush the tubes so that divisions or markings can be clearly seen at all times.

A manually operated admixture dispenser system shall be furnished a valve with a locking mechanism to prolong the discharge cycle until it is ascertained that all admixture has been discharged.

When finely divided mineral admixtures are used in bulk, the weighing sequence shall be cement first and admixture or pigment second, followed by aggregates. This procedure shall be followed when cement and finely divided mineral admixtures are weighed cumulatively on the same scale beam. Pigments shall be preweighed and batched from packages that are of a size appropriate for a single batch. Mineral admixtures shall not be charged into a wet mixer ahead of the other materials. Likewise, they shall not be charged into the mixer at the same time as the mixing water.

With central mixed concrete, silica fume shall be added after (or if ribbon fed, along with) all other ingredients. For truck mixed concrete, the silica fume shall be added to the truck before any other ingredients since it contains much of the batch water needed for the mix. Metakaolin shall be added to the batch along with the cement or added after all other ingredients. Care shall be taken when adding metakaolin after all other ingredients to ensure that the mixer is not overloaded and the slump of the concrete is greater than 4 in. (100 mm).

Commentary

If more than one admixture is being used through a single dispenser without flushing of the dispenser with water after each cycle, it is necessary to ascertain that the admixtures are compatible and that the mixing of the admixtures prior to introduction in the mix will not be detrimental. An adverse interaction might interfere with the efficiency of either admixture or adversely affect the concrete.

Although admixture batching systems usually are installed and maintained by the admixture producer, plant operators should thoroughly understand the system and be able to adjust it and perform simple maintenance. Prior to installation of the dispenser, the system should be analyzed carefully to determine what possible batching errors could occur and, with the help of the admixture supplier, they should be eliminated.

Mineral admixtures have a tendency to stick to the sides of a wet mixer drum when charged ahead of other materials. Also, they have a tendency to ball up when charged into the mixer at the same time as the mixing water.

Adding silica fume last to a truck mixer may result in "head packing" and inadequate mixing, particularly in larger loads. Adding dry densified silica fume last to a truck mixer may produce inadequate mixing, particularly in larger loads.

Standard

4.17 Mixing of Concrete

4.17.1 General

All mixers and methods used shall be capable of thoroughly combining the concrete materials containing the largest specified aggregate into a uniform mass of the lowest slump practical for the product being produced within the established mixing time or number of revolutions.

Care shall be taken to ensure that the weighed materials are properly sequenced and blended during charging of the mixers. All concrete materials shall be discharged to the mixer while the drum or blades are rotating. In the discharge of component materials from batching facilities to the mixer, the solid materials shall, if possible, be arranged in the charging hoppers so proportional amounts of each will be ribbon fed to the mixer.

4.17.2 Methods of Concrete Mixing

Concrete shall be mixed by one of the following methods:

1. Central mixed concrete – concrete mixed in a central stationary mixer and delivered to the casting area by buckets, truck agitator or non-agitating trucks.

2. Shrink mixed concrete – concrete that is partially mixed in a stationary mixer, then mixed completely and delivered to the casting site in a truck mixer.

3. Truck mixed concrete – concrete that is completely mixed in a truck mixer as it is delivered to the casting site (for backup mixes only)

All requirements for mixing of concrete as given in this Article are valid for both normal weight aggregates as specified in ASTM C33 and for lightweight aggregates for structural concrete as specified in ASTM C330.

The time from start of mixing to placement shall not exceed 1 hour. Retempering (with water) of concrete which has started to stiffen shall not be allowed.

When colored concrete, or buff or white cement mixes are used in conjunction with gray concrete—e.g., with facing and backup mixes,—sep-

Commentary

C4.17 Mixing of Concrete

C4.17.1 General

Concrete of satisfactory quality requires the materials to be thoroughly mixed until there is a uniform distribution of the materials and the mix is uniform in appearance. The necessary mixing time will depend on many factors including batch size, workability of the batch, size and grading of the aggregate, type of mixer, condition of blades, and the ability of the mixer to produce uniform concrete throughout the batch and from batch to batch. With some mixes, to reduce the potential of surface voids (bugholes), the concrete should be mixed slightly longer than normal to help break up water droplets and large air pockets and at the same time produce better workability in the concrete mass.

Varying the mixing time from batch to batch may cause different degrees of dispersion of cement or coloring pigment, if used, and therefore, different shades of color. This is particularly important when using white cement with or without color pigments and when the colors of the sand and cement are different.

C4.17.2 Methods of Concrete Mixing

3. Truck mixing is only suitable for back-up mixes because it usually results in insufficient uniformity for uniform visual appearance in architectural units.

The practice of adding superplasticizer to counteract slump loss must be carefully monitored to avoid potential uniformity problems.

Standard

arate mixers and handling arrangements are required. Alternatively, the equipment shall be water flushed several times and completely cleaned to remove all concrete residue before being used for producing a different color mix.

4.17.3 Mixing Time and Concrete Uniformity

Mixing time shall be measured from the time all cement and aggregate are in the mixer drum. All water shall be in the drum by the end of the first one-fourth of the established mixing time. Mixing time for air-entrained concrete shall be verified and controlled to ensure the specified air content. The mixing time required for each batch shall be uniform and based upon the ability of the mixer to produce uniform concrete throughout the batch and from batch to batch. For each type of mixer, the optimum mixing time shall be established by manufacturer's recommendations, or mixer performance shall be determined from tests using two samples, one taken after discharge of approximately 15 percent and the other after 85 percent of the batch, but not more than 15 minutes apart, see Table 4.17.3.

When uniformity sampling is done, slump tests of individual samples taken at the time of sampling shall be made for a quick check of the probable degree of uniformity. If the slumps differ more than that specified in Table 4.17.3 the mixer shall not be used unless the condition is corrected or the operation changed to a longer mixing time, a smaller load, or more efficient charging sequence that will permit the requirements of Table 4.17.3 to be met.

Table 4.17.3. Requirements for uniformity of concrete (from ASTM C-94)

Test	Maximum Allowable Deviation Between Samples
Weight per cubic foot calculated to an air-free basis, lb/ft^3	1.0
Air content, volume % of concrete	1.0
Slump: If average slump is 4 in. or less, in. If average slump is 4 to 6 in. If superplasticizers used in mix, in.	1.0 1.5 2.5
Coarse aggregate content, portion by weight of each sample retained on No. 4 sieve, %	6.0
Unit weight of air-free mortar based on average for all comparative samples tested, %.	1.6

Commentary

C4.17.3 Mixing Time and Concrete Uniformity

Both overmixing and undermixing are to be avoided. Under mixing will result in concrete of variable consistency and low strength. Overmixing results in loss of air in air entrained mixes, grinding of aggregates, and loss of workability. Concrete mixing procedures should be established for each type of mixer. Variations in mix designs such as lightweight concrete and the use of superplasticizers may require adjustments to standard mixing procedures used for other mixes.

The amount of air entrained varies with the type and condition of the mixer, the amount of concrete being mixed, and the mixing speed and duration. Air content increases as mixer capacity is approached. More air is entrained as the speed of mixing is increased up to 20 rpm and decreases at higher speeds. The amount of entrained air increases with mixing time up to about 5 minutes, beyond which it slowly decreases. However, the air void system, as characterized by specific surface and spacing factors, generally is not harmed by prolonged agitation.

Mixers should not be loaded above their rated capacities and should be operated at the speeds for which they were designed. It may prove beneficial to reduce the batch size below the rated capacity to ensure more efficient mixing. Increased output should be obtained by using a larger mixer or additional mixers, rather than by speeding up or overloading the equipment on hand. If the blades of the mixer become worn or coated with hardened concrete, the mixing action will be less efficient. Badly worn blades should be replaced and hardened concrete should be removed periodically, preferably after each day's production of concrete.

The most positive control method for maintaining batch-to-batch uniformity is a regularly scheduled program of uniformity tests of the fresh concrete. Scheduling of operations should be carefully controlled to prevent any delays in the period between charging the mixer and depositing the concrete in the molds. Such delays, which cause the concrete to be held in mixers, transporting equipment or buckets, or anywhere else tend to cause non-uniformity in the placed concrete.

See ACI Manual of Concrete Inspection SP-2 for procedures to determine coarse aggregate content and air-free unit weight of mortars.

Standard

4.17.4 Mixing Time - Stationary Mixers

Unless otherwise recommended by the mixer manufacturer, for stationary mixers, the minimum mixing time shall be one minute for batches of 1 cu yd (0.8 m³) or less. This mixing time shall be increased by at least 15 seconds for each cubic yard, or fraction hereof, of capacity in excess of 1 cu yd (0.8 m³). The mixing time shall be increased when necessary to secure the required uniformity, particularly for colored concrete, and consistency of the concrete, but mixing time shall not exceed three times the specified time. In case a batch has to be held in the mixer for a longer period, the speed of the drum or blades shall be reduced to an agitating speed or the mixer stopped for intervals to prevent over mixing. Both undermixing and overmixing shall be avoided.

4.17.5 Mixing Time - Shrink Mixing

When a stationary mixer is used for partial mixing of the concrete, the time may be reduced to a minimum of 30 seconds, followed by not less than 50 or more than 100 revolutions at mixing speed in a truck mixer. Any additional revolutions in the truck mixer shall be at agitating speed in accordance with mixer manufacturers recommendations.

4.17.6 Mixing Time - Truck Mixing

For mixing in a truck mixer loaded to its maximum capacity, the number of revolutions of the drum or blades at mixing speed shall be not less than 70 nor more than 100. If the batch is at least 1/2 cu yd (0.4 m³) less than the maximum capacity, the number of revolutions at mixing speed may be reduced to 50. All revolutions in excess of 100 shall be at agitating speed per mixer manufacturer's recommendations.

An absolute minimum of 100 revolutions at a speed of at least 15 revolutions per minute shall be used for truck mixers mixing concrete containing silica fume or metakaolin. When using dry densified silica fume, these requirements must be increased to 120 revolutions at a minimum speed of 15 revolutions per minute. The load size shall be limited to not more than 75 percent of rated mixing capacity when truck mixing mixes including silica fume.

4.17.7 Special Batching and Mixing Requirements for Lightweight Aggregates

Lightweight aggregate concrete shall be batched and mixed as recommended by the producer of the aggregate. Batch weights of lightweight aggregates shall be based on oven-dry weights corrected for ab-

Commentary

C4.17.4 Mixing Time - Stationary Mixers

Mixing time less than that specified by the manufacturer may be permitted provided performance tests indicate that the time is sufficient to produce uniform concrete.

C4.17.7 Special Batching and Mixing Requirements for Lightweight Aggregates

ACI 211.2 Standard Practice for Selecting Proportions for Structural Lightweight Concrete provides additional information on the use of lightweight aggregates.

When a lightweight concrete mix is used for the first time, the

Standard

sorbed moisture or absorbed moisture plus surface moisture.

Lightweight aggregate may require pre-dampening prior to batching. Aggregate shall be tested for water absorption with the minimum moisture content likely to occur during production.

With aggregates having less than 10 percent total absorption by weight or shown to absorb less than 2 percent water by weight during the first hour after immersion in water, pre-dampening prior to mixing is not required.

If lightweight aggregate absorbs more than 2 percent water by weight, it shall be pre-dampened and a mixing procedure developed that is shown to produce concrete of uniform quality.

4.17.8 Cold Weather Mixing

Concrete temperatures at the mixer shall be maintained above a minimum of 50 deg. F (10 deg. C). Materials shall be free from ice, snow and frozen lumps before entering the mixer.

When exposure to cold weather is severe, due either to low air temperature, or because the concrete sections are thin, the temperature of concrete shall be increased by heating to assure the concrete temperature does not fall below 50 deg. F (10 deg. C). To accomplish this, the mixing water and, if necessary, the fine aggregates shall be heated. Aggregates shall not be heated above 180 deg. F (80 deg. C).

To avoid the possibility of premature stiffening or flash set of the cement when either the water or aggregates are heated to a temperature over 100 deg. F (40 deg. C), water and aggregate shall come together first in the mixer so that the temperature of the combination is reduced below 100 deg. F (40 deg. C) before the cement is added.

Any material used in a liquid or slurry form shall be protected from freezing. Powdered materials shall be protected from moisture. Any material that has deteriorated, has been damaged by frost, or has

Commentary

mixing and handling procedures should be checked to make sure they are adequate to produce uniform concrete of the quality specified. In such a case, the aggregate producer's recommendations may be particularly helpful.

Knowledge of the aggregate water absorption characteristics simplifies the process of batching lightweight mixtures, assuring proper mixing in a minimum of time. If lightweight aggregate unit weight varies, then batching by volume rather than by weight is necessary.

C4.17.8 Cold Weather Mixing

To help provide for products of uniform color, concrete temperatures should be as uniform as practicable from batch to batch.

When concrete temperatures are less than 50 deg. F (10 deg. C) the time required for the concrete to gain strength is greatly extended or stopped.

Where hot water is used for maintaining a minimum concrete temperature, provisions should be made for the operator to read the temperature of the water before it enters the mixer and after possible blending with cold water.

Standard

been contaminated shall not be used in the production of concrete.

4.17.9 Hot Weather Mixing

Concrete temperature at the mixer shall be maintained below a maximum of 95 deg. F (35 deg. C)

If high temperatures are encountered, the ingredients shall be cooled before or during mixing, or flake ice or well-crushed ice of a size that will melt completely during mixing, may be substituted for all or part of the mixing water to avoid low slump, flash set or cold joints.

4.18 Requirements for Transporting and Placing of Concrete

4.18.1 General

Proven and effective procedures for placing concrete are described in detail in the following publications of the American Concrete Institute:

1. Recommended Practice for Measuring, Mixing, Transporting, and Placing Concrete (ACI 304)
2. Recommended Practice for Hot Weather Concreting (ACI 305).
3. Recommended Practice for Cold Weather Concreting (ACI 306)

These publications shall be available in precast concrete plants and supervisory personnel shall be familiar with their contents.

Sufficient mixing and placing capacity shall be provided so the products are free from unplanned cold

Commentary

C4.17.9 Hot Weather Mixing

Care should be taken in hot climates to protect batch material storage bins and water lines from direct sun.

Concrete temperatures higher than 80 deg. F (27 deg. C) will result in a faster setting rate, visible flow lines, and possible cold joints. Since high temperature concrete has a reduced workability time available for placing and consolidating, proper scheduling of the concrete placement is necessary.

High temperatures in face mixes at time of placing can create unacceptable finishes and variations in color.

If false set occurs, the mix should sit for 1 to 3 minutes and then the mixer restarted for 30 additional seconds. The remixing of false set will mechanically break the false set and provide a usable mix.

C4.18 Requirements for Transporting and Placing of Concrete

C4.18.1 General

In arranging equipment to minimize separation or segregation, it is important that the concrete drop vertically into the center of whatever container receives it. The importance of this increases greatly with increases in slump, maximum size and amounts of coarse aggregate, and with reduction in cement content. The height of free fall of concrete need not be limited unless a separation of coarse particles occurs (resulting in honeycomb) or uniformity of appearance is affected, in which case a limit of 3 to 5 ft (0.9 to 1.5 m) may be adequate. However, to protect spacers, embedded features and mold surfaces, and to prevent displacement of reinforcement, concrete fall in molds should be limited, by means of a suitable drop chute or other device, to within a few feet of concrete in place. Whether such a fall is into mold or into a hopper or bucket, the final portion of the drop should be vertical and without interference. If molds are sufficiently open so concrete can be dropped into them with little or no disturbance to reinforcement or face mix, etc., direct discharge into molds is usually faster and saves the labor of placing, moving and removing hoppers and drop chutes. Where direct dumping is practicable, it results in little but unimportant scattered separation, whereas fixed drop chutes often cause considerable separation due either to flowing or stacking of the concrete.

If concrete is placed in layers to produce a monolithic and visually acceptable finished placement, it is important each layer

Standard

joints or a plane of separation. If delays occur which result in the concrete attaining its initial set so it will not receive a vibrator and again become plastic, partially filled molds shall be washed out or partially cast units rejected.

The concrete consistency and desired surface appearance will also affect the placement method employed; the lower the slump or water-cement ratio, the shallower the lift that shall be used. The time lapse between placing face mix and backup mix shall be short enough to ensure proper plastic bonding between the layers and shall occur prior to face mix attaining its initial set.

4.18.2 Transporting and Placing Concrete

Since the workability of concrete decreases with time after mixing, the concrete shall be transported from mixer to forms in the shortest possible time and in such a manner as to prevent segregation or loss of mortar.

Free water or excess grout shall not drip from buckets onto a retarder film or previously finished concrete during the placing operation.

Before the beginning of a new run of concrete, hardened concrete and foreign matter shall be removed from the surfaces of the equipment involved.

The need for agitation during transportation will depend on the length of time between completion of mixing and discharge of the concrete.

Regardless of the manner of transportation, concrete shall be discharged into the molds while in its original mixed or plastic state. Retempering by adding water and remixing concrete which has started to stiffen shall not be permitted.

4.18.3 Preventing Aggregate Segregation

Only those methods and arrangements of equipment shall be used which result in placing concrete in a uniform, compacted condition without separation of coarse aggregate and paste. Obvious clusters and pockets of coarse aggregate are objectionable and shall be scattered prior to placing concrete over them to prevent rock pockets and honeycomb in the completed unit. Placing methods used shall preserve the quality of the concrete, in terms of water-cement ratio, slump, air content, and homogeneity.

Commentary

of concrete be shallow enough so it may be placed while the previous layer is still plastic (prior to initial set and/or evaporation of bleed water) and the two layers be vibrated together.

C4.18.2 Transporting and Placing Concrete

Transportation of concrete to the placing site should be evaluated and procedures maintained to avoid unnecessary changes to the concrete.

The use of non-agitating trucks to transport concrete containing silica fume or metakaolin is not recommended.

C4.18.3 Preventing Aggregate Segregation

Scattered individual pieces of separated coarse aggregate are not objectionable if they are readily enclosed and consolidated into the concrete as it is placed and vibrated.

It is a common fallacy to believe that segregation occurring in handling will be eliminated by subsequent placing and consolidation operations.

Concrete should not be dumped in separate piles and then leveled and worked together; nor should the concrete be deposited in large piles and moved horizontally into final position. Such practices result in segregation because mortar tends to flow ahead of coarser material with resultant visible flow lines on exposed surface. Also, placing concrete on concrete already in the mold minimizes loose aggregate rollout and swirl patterns during consolidation, and it minimizes additional entrapped air

Standard

4.18.4 Preparation of the Molds

Prior to concreting, the mold shall be cleaned and any unwanted materials, such as tie wire clippings, shall be removed by compressed air, vacuum cleaning, or other methods. If required, before placing either reinforcing steel or concrete, a form release agent or retarder shall be carefully and uniformly applied to all mold surfaces that will be in contact with the concrete to provide easy release and stripping of the element from the mold consistent with the required finish.

4.18.5 Placing Concrete Under Severe Weather Conditions

Freshly deposited concrete shall be protected from freezing, excessively high or differential temperatures, premature drying, and moisture loss for a period of time necessary to develop the desired properties of the concrete.

Concrete as placed shall be at a temperature between 50 deg. F (10 deg. C) and 95 deg. F (35 deg. C) and as uniform as is practical from one batch of concrete to another for a specific project. Measures shall be taken during severe weather concreting operations so that the concrete remains plastic for a suitable placing and finishing period and then gains strength under favorable curing conditions.

4.18.6 Placing Concrete in Wet and Rainy Conditions

Because of the nature of the architectural precast concrete production process, the producer shall have adequate weather protection provisions on hand at all times for outside production activities. It shall be possible to deploy the weather protection provisions without compromising the quality of the product.

Commentary

between the deposits at the mold face.

Where mixtures of dry or stiff consistencies are required, the placement rate should be slow enough to permit proper vibration in order to avoid bugholes and honeycombing. Placing concrete too slowly can produce layer lines or cold joints due to improper consolidation. The rate of placement and vibration factors (intensity, spacing) should be selected to minimize entrapped air in the concrete. Thin, even layers of 6 in. (150 mm) or less can be consolidated with only small air holes likely to show; layers thicker than 12 in. (300 mm) will increase the possibility of trapping more and larger air holes. For this reason, the use of a face mix or a specially placed mix of only 1 to 2 in. (25 to 50 mm) thick may be advisable.

C4.18.5 Placing Concrete Under Severe Weather Conditions

Using weather data for the production area, temperature extremes and durations can be identified. A plan of procedures for severe weather conditions should be developed for use whenever weather conditions dictate.

Optimum concrete temperature for placing is recommended as 70 deg. F (20 deg. C).

Standard

4.18.7 Placing Concrete in Hot or Windy Conditions

For concrete placed in hot weather, the temperature of concrete at time of placing shall not be above 95 deg. F (35 deg. C). Adequate moisture shall be retained in the concrete during the curing period to prevent surface drying. The temperature of the concrete as placed shall not be so high as to cause difficulty from loss of slump, flash set, or cold joints.

When the ambient temperature is above 80 deg. F (27 deg. C), methods shall be taken to protect concrete in place from the effects of hot and/or drying weather conditions.

Commentary

C4.18.7 Placing Concrete in Hot or Windy Conditions

The upper limit of 95 deg. F (35 deg. C) for concrete placement temperature is a guideline which may need to be lowered for specific materials which respond negatively at higher temperatures. If problems are encountered using 95 deg. F (35 deg. C) as an upper limit for materials in use, the maximum temperature should be reduced until problems are eliminated.

Wind or direct sunlight can have significant and rapid drying and shrinkage effects. These along with low humidity are items which must be considered and adjusted for during placement of concrete.

If hot and/or drying weather conditions occur, any of the following procedures or combination of procedures may be used to prevent plastic shrinkage cracking of concrete or loss of strength:

1. Shaded storage for aggregates.

2. Sprinkling or fog spraying of aggregates or chilling aggregates with liquid nitrogen

3. Burying, insulating and/or shading water-supply facilities.

4. Use of cold water in batching.

5. Use of shaved or crushed ice for a portion of the mixing water. Only as much ice should be used as will be entirely melted at the completion of the mixing period.

6. Maintaining concrete surfaces in a cool and moist condition by use of coverings such as wet burlap or by sprinkling or ponding as soon as exposed concrete surfaces are finished, or as soon as water sheen disappears. This is especially important for hot, windy, exposed locations.

7. Shading of product surface during and after casting to avoid heat buildup in direct sunlight.

8. Do not use cement with temperatures over 170 deg. F (77 deg. C)

9. White pigmented membranes may be used but are not recommended in very hot weather for the first 24 hr. after casting since they do not cool the concrete as much as wet curing methods.

10. A set-retarding admixture included in the concrete mix may delay concrete setting time in hot weather and give a longer period for placing and finishing.

11. When the temperature of steel is greater than 120 deg. F (50 deg. C), steel molds and reinforcement should be sprayed with water just prior to placing the concrete. Surface of mold should be free of visible water droplets prior to placing concrete since water droplets will often appear as voids on the concrete surface, even if properly vibrated.

Standard

4.18.8 Placing Concrete in Cold Weather Conditions

Special precautions shall be taken to protect concrete placements in cold weather [below 40 deg. F (5deg. C)] so that the concrete gains strength under favorable curing conditions.

4.18.9 Placing Facing Concrete

When placing the initial concrete face mix, care shall be taken to avoid coating the reinforcement with cement paste which may endanger proper bonding in the second stage of concreting. Where this coating cannot be prevented, the placing of the backup concrete mix shall be done prior to the hardening of such coating, or it shall be cleaned off. Alternatively the placement of the reinforcement shall follow the consolidation of the face mix.

Facing concrete shall be carefully placed and worked into all details of the mold. This is particularly important in external and internal corners for true and sharp casting lines. Concrete shall be placed so that flow is away from corners and ends rather than toward them. Each batch of concrete shall be carefully ribboned on, as nearly as possible, in final position, into previously cast concrete until the mold is filled and the whole mass consolidated by vibration with as little shifting as possible. Concrete shall be spread, if necessary, with a short-handled, square-end shovel, a come-along, hoe, or similar solid faced tools. If toothed rakes are used to spread concrete, care shall be taken to prevent concrete segregation

Concrete shall not be subjected to any procedure which will cause segregation or flow lines. When a

Commentary

C4.18.8 Placing Concrete in Cold Weather Conditions

Concrete can be placed successfully when ambient temperature falls below 40 deg. F (5 deg. C). This is done by heating the molds to maintain a minimum concrete temperature of 50 deg. F (10 deg. C). The existence of windy and cold conditions may dictate added protection to maintain temperature of the mold and the concrete after placement.

The mold temperature should be about the same as the concrete temperature. An optimum of 75 to 80 deg. F (21 to 27 deg. C) is desirable. It is more important that the temperatures remain relatively constant from unit to unit throughout the project rather than whether they are near the upper or lower level of a temperature range.

C4.18.9 Placing Facing Concrete

Paste coating of strand or mild steel reinforcement during placement should be of no concern up to the top of the section; above that, paste should be kept from or cleaned from all reinforcement.

The use of separate face and backup mixes or the use of a uniform concrete mix throughout the unit depends on economics, the type of finish required, the configuration of the unit and the practice of the particular precasting plant. Members with intricate shapes and deep narrow sections generally require one uniform concrete mix throughout.

Where an aggregate or finished face material changes within a panel, a definite feature (demarcation strip) should be inserted into the panel face to facilitate casting and to achieve an acceptable demarcation joint between finishes. Two concrete mixes with different colored matrices exposed at the face of the panel should generally not be used within one unit unless a technique is used to prevent mixing of the two mixes. For example, the two-stage (sequential) precasting method where one part of a unit is cast first of one mix and, after curing, is cast into the total unit has been successfully used.

Units with large or steep returns (such as channel column covers and some spandrels) may be cast in separate pieces in order to achieve matching high quality finishes on all exposed faces and then joined with dry joints. Although the dry joint may not show with certain mixes and textures, a groove or quirk will help to mask the joint. Where desired, this joint can be recessed deep enough to allow installation of a small backer rod and placement of a 1/4 in. (6 mm) bead of joint sealant. Sometimes precautions may be necessary to ensure water tightness of the dry joints. Care should be taken to ensure that the return is jigged and held securely at the proper angle during production of the base panel.

Standard

retarder is used, the concrete shall not be moved over the mold in such a manner that it would disturb the retarder. Concrete placing on a slope shall begin at the lower end of the slope and progress upward thereby increasing compaction of the concrete.

The thickness of a face mix after consolidation shall be at least 1 in. (25 mm) or a minimum of 1-1/2 times the maximum size of the coarse aggregate, whichever is larger. A face mix shall be thick enough to prevent any backup concrete from showing on the exposed face.

4.18.10 Placing Backup Concrete

In placing backup concrete, care shall be taken to break the fall into the mold so that it will not displace the face mix.

4.19 Consolidation of Concrete

4.19.1 General

Concrete used in architectural precast concrete units may be uniformly consolidated by internal, or external vibration (with or without vibrating table), surface vibrators, spading, by impact, or a combination of these methods.

Consolidation of concrete mixes shall accomplish full coating of the coarse aggregate and reinforcement with cement paste. Air pockets around uncoated reinforcement which create potentially corrosive environments shall be avoided by the selected consolidation methods.

After the proper vibrating equipment has been selected, it shall be operated by conscientious and well-trained operators who will consistently maintain the proper vibrator spacing and vibration times, and who have been trained to judge when the concrete is adequately consolidated.

Regardless of the techniques used for vibration, its effectiveness shall be judged by the surface condition of the finished concrete, unless circumstances indicate a more in-depth evaluation is needed. If surface defects such as honeycombing, aggregate or mortar pockets and excessive air bubbles are common, the vibration procedure, number and depth of lifts or other procedures shall be revised as necessary to produce acceptable surfaces.

Vibration procedures shall be evaluated at the beginning of a project to determine the vibration time for each type of vibrator for each mixture, mold

Commentary

The 1 inch (25mm) minimum thickness of face mix dimension is chosen because the consolidated face is normally used to provide the proper concrete cover over the reinforcement. For units not exposed to weather or for face mixes applied face-up, this dimension may be reduced to 3/4 in. (20 mm) provided the backup mix does not bleed through the face mix.

C4.18.10 Placing Backup Concrete

The allowable time interval between placing face mix and backup mix depends on the concrete mix, the temperatures of the mix and the ambient air, and the drying conditions near the mold.

C4.19 Consolidation of Concrete

C4.19.1 General

Vibration should be distributed so the concrete is thoroughly consolidated, producing a dense, uniform mass with surfaces free of imperfections or blemishes. The optimum time of vibration depends on the type of vibrator, the mix characteristics, and the configurations of mold and reinforcement. Reducing the vibration time on the last lift in returns will result in increased bugholes.

The selection of the best vibrator or vibration method involves factors such as:

1. Size, shape, type and stiffness of molds.
2. Concrete mix and consistency.
3. Plant preference based on experience.

Unless there is a gross violation of recommended mix proportioning and vibration procedures, it is difficult to harm concrete by excessive vibration, provided that the forms are designed to withstand the longer durations of high frequency vibration. Under vibration is far more common than over vibration.

Standard

type and configuration. Each member and concrete layer shall be vibrated for the same time duration, consistent with the slump of the concrete as delivered to the mold.

4.19.2 Consolidation of Lightweight Concrete

In consolidation of lightweight concrete precautions shall be taken to avoid aggregate floatation caused by over vibration.

4.19.3 Consolidation of Face and Backup Mixes

Special attention is required for consolidation of face mixes, especially if the aggregates are to be exposed later by removing the paste from the surface.

The layers of face mix concrete shall be as level as possible so that the vibrator does not need to move the concrete laterally, since this will cause segregation, see Article 4.18.9. Where there are some mounds or high spots in the surface of the concrete as placed, a vibrator shall not be stuck into the center of each mound to knock it down, but instead leveling shall be by hand or mechanical screed.

4.19.4 Use of Internal Vibrators

Care shall be taken to ensure proper vibration with minimal penetration of the backup mix into the face mix. All vibrating shall be done by workers specifically trained to use internal vibrators in the correct manner, with emphasis on the fact that vibrators in concrete must be moving at all times.

Commentary

C4.19.2 Consolidation of Lightweight Concrete

In consolidation of lightweight aggregate concrete, care should be taken not to over vibrate. Since the coarse aggregate particles are the lightest solid ingredients in the mix, over vibration can cause the particles to rise, so that finishing problems result from "floating" aggregate which could cause the strength to be non-uniform through the depth of the member. In addition to care in vibration, the use of slumps less than 4 in. (100 mm) helps greatly in avoiding segregation, during both handling of the mix and consolidation and finishing.

C4.19.3 Consolidation of Face and Backup Mixes

During vibration each individual coarse aggregate particle moves in all directions through the mortar. At the same time each particle is turning about its shorter axis and striving to occupy a position so that it can penetrate the cement mortar most easily. As the aggregate penetrates and thoroughly mixes with the mortar, any water or air bubbles that are trapped beneath it become released and start to move upward. The process can occur only if there is sufficient mortar present.

Experienced and competent vibrator operators working with regularly maintained vibrators and sufficient standby units are essential to a satisfactory consolidation program. There is a tendency for inexperienced vibrator operators to merely flatten the batch. Proper consolidation is assured only when the other items evidencing adequate vibration are sought and attained.

C4.19.4 Use of Internal Vibrators

The distance between vibrator insertions shall generally be about 1 to 1-1/2 times the radius of action, or such that the area visibly affected by the vibrator overlaps the adjacent just-vibrated area by a few inches. The vibrator shall not be inserted within 2 ft. (0.6 m) of any leading (unconfined) edge.

Employing internal vibrators in this manner will minimize entrapped air between the concrete and the mold and blend the two layers.

Consolidating Flowing Concrete. When casting units with constricted dimensions or limitations on the amount of consolidation effort which may be applied, a high slump, flowing concrete should be used. When consolidating this concrete, use a large amplitude (i.e., large diameter) internal vibrators inserted at a close spacing and withdrawn slowly. While consolidating in this manner, the surface should be examined for evidence of excess water or paste, and if this does appear the amount of consolidation effort used should be reduced. The high-slump, self-leveling characteristics of the flowing concrete may appear not to need any consolidation effort. However, at least nominal vibration should be provided to eliminate large air voids.

Standard

Vibrators shall not be allowed to contact forms for exposed concrete surfaces. Internal vibrators shall not be inserted closer to the form than 2 to 3 in. (50 to 75 mm). Care shall be taken to avoid vibrating the reinforcement. Also, care shall be taken not to displace cast-in hardware or disturb the face mix.

When internal vibrators are used to consolidate concrete around epoxy-coated steel reinforcing bars, the vibrators shall be equipped with rubber or nonmetallic vibrator heads.

4.19.5 Use of External Form Vibrators

External vibrators shall be securely fixed to the forms to obtain the maximum vibration effect and to avoid damage to vibrators and forms. During the casting operation, a check shall be made to verify that the vibrators are in operation and firmly in place.

The size and spacing of form vibrators shall be such that the proper intensity of vibration is distributed evenly over the desired area of form and no form damage occurs.

4.19.6 Use of Surface Vibrators

Surface vibrators shall be of a type and shape, and shall be applied in such a manner that will prevent suction action during use. Surface vibrators shall be moved at a rate so vibration is sufficient to embed the coarse aggregate and bring enough paste to the surface for finishing. The vibration and rate of movement shall be sufficient to compact the full depth of the concrete layer. Solid plate and bull float vibrators shall be operated at a slight angle from the plane of the concrete surface for best results.

If grate tampers are used, the concrete slump shall not be over 2 in. (50 mm). Vibrating grate tampers shall not be used for structural lightweight aggregate concrete.

Commentary

The vibration frequencies which affect the aggregates and fines of a concrete mix range from 1,200 to 14,000 cycles per minute. The lower frequencies activate the large diameter particles, and the higher frequencies affect the fines and cement in the mix and assure coverage of larger particles with a complete film of grout.

Vibrators contacting forms for exposed concrete surfaces may mar and disfigure the concrete surface. Internal vibrators inserted close to the form may cause the coarse aggregate to be driven away from the form face producing local pockets of fine material (stinger marks) on the surface. Vibration of reinforcement may cause reinforcing steel reflection features visible in the surface of the finished product.

C4.19.5 Use of External Form Vibrators

The spacing of external form vibrators is a function of the type, stiffness and shape of the form, depth and thickness of the concrete, force output per vibrator, workability of the mix, and vibrating time.

Forms should be placed on neoprene isolation pads or other resilient base material to prevent transmission and loss of vibration to the supporting foundation.

On large projects trial production of typical pieces may be appropriate as present knowledge is inadequate to predict an exact solution to the complex problem of vibrator spacing.

C4.19.6 Use of Surface Vibrators

Surface vibrators which may be used to consolidate thin layers of concrete like facing mixes include vibrating screeds; a small plate or grid vibratory tamper [usually 2 to 3 ft (0.6 to 0.9 m) square in area] powered by an external air vibrator; hand jitterbug; bull floats with vibrators mounted along the rib section of the floats. Vibrators on the bull floats should be mounted so that their shafts rotate in opposite directions.

When using surface vibrators, care should be exercised to prevent a smooth face mix back surface which may result in poor bond of backup concrete.

Standard

4.19.7 Use of Vibrating Tables

Care shall be taken to obtain the proper distribution of vibration. The number and location of external vibrators to be used on a vibrating table shall be determined on the basis of adequate amplitudes of vibration and uniform distribution over the entire concrete surface. The frequency and amplitude shall be checked at several points on the table, using a vibrograph, vibrating reed tachometer, or other suitable method. The vibrators shall be moved around until dead spots are eliminated and the most uniform vibration is attained.

4.20 Requirements for Curing Concrete

4.20.1 General

Freshly deposited and consolidated concrete shall be protected from premature drying and extremes of temperature. The curing concrete shall be maintained with minimal moisture loss at a relatively constant temperature for the period of time necessary for the hydration of the cement and proper hardening of the concrete. Curing procedures shall be well established and properly controlled to develop the required concrete quality and stripping or transfer strength. Curing procedures shall minimize any appearance blemishes, such as non-uniformity of color, staining or plastic shrinkage cracking.

To produce concrete of uniform appearance, consistent and uniform curing conditions shall be provided.

Protection of concrete surfaces against moisture loss to prevent shrinkage cracking of silica fume or metakaolin concrete shall begin immediately after finishing, whatever the finishing process may be. High dosages of silica fume or metakaolin produce concrete with significantly reduced bleeding. Therefore, there is no requirement to wait for the bleeding to conclude before initiating protection of surface.

4.20.2 Curing Temperature Requirements

The concrete in the mold shall be maintained at a temperature of not less than 50 deg. F (10 deg. C) during the curing period. Curing shall be started immediately following initial set of the concrete. The time between placing of concrete and the start of curing shall be minimized in hot or windy weather to prevent loss of moisture, and the resultant drying and plastic shrinkage cracking.

Commentary

C4.19.7 Use of Vibrating Tables

Vibrating tables or casting decks are best used for flat or low profile units, and provide an easy and effective method for application of external vibration

The frequency and amplitude of vibrating forms and vibrating tables equipped with external vibrators should be determined at sufficient points to establish their distribution over the surface. Inadequate amplitudes cause poor consolidation, while excessive local amplitudes cause the concrete to roll and tumble so that it does not consolidate properly.

C4.20 Requirements for Curing Concrete

C4.20.1 General

The curing of concrete depends on many variables, including the mass of the member, type and properties of cement, air temperature, humidity and other variables. The curing period of concrete which is of significant interest is during the early stages of strength development from initial set until the concrete has reached a design strength for stripping or stressing the member. Curing involves the maintaining of a satisfactory moisture content and temperature in concrete. If concrete is allowed to lose moisture, it may suffer a lack of strength development and potential plastic shrinkage cracking as the surface moisture is lost. Therefore, concrete surfaces should be protected to prevent rapid loss of moisture while the concrete is plastic.

C4.20.2 Curing Temperature Requirements

Standard

During the initial curing period (prior to stripping) positive action shall be taken to provide heat (if necessary to maintain minimum temperatures) and to prevent loss of moisture from the unit. Curing materials or methods shall not allow one portion of an element to cure differently or dry out faster than other portions of the element.

4.20.3 Curing to Attain Specified Stripping or Release Strength

Curing temperature and duration of curing shall assure the stripping strength indicated on the production drawings has been reached before a product is removed from its mold. The stripping strength shall be set by the design engineer, based on the characteristics of the product. It shall be high enough to ensure that the stripping does not have a deleterious effect on the performance or the appearance of the final product.

Prior to stripping, transfer of stress, or handling, the strength of the product shall be determined by test specimens cured, as nearly as possible, in the same manner as the product.

In addition to regular test cylinders cured in accordance with ASTM requirements, for units where shipping or erection strength may be critical, extra test cylinders shall be made and cured similarly to the precast concrete units to verify curing results and estimate the strength of units at time of shipping.

The establishment of a strength level where exposure to local climatic conditions is considered safe by individual plant standards or local practices (see Article C4.2.) shall not relieve the precast concrete manufacturer of responsibility to verify that the products have reached the required final design strength.

4.21 Accelerated Curing of Concrete

4.21.1 General

Accelerated curing procedures shall be developed based on efficiency of concrete strength development without damaging the concrete. Temperature guidelines for accelerated curing, no matter which method is used, are as follows:

1. The controlling temperatures shall be those achieved within the concrete elements, not

Commentary

Retention of the heat released by the hydration of the cement can be used advantageously by precasters to provide much of the heat for curing. Insulated tarpaulins are effectively used for a combination of moisture and heat retention. Accelerated curing is the use of added heat to increase the rate of the hydration reactions.

Differential curing of an element may produce color variations in the finished product and differential shrinkage of the concrete which can lead to warping of the element.

C4.20.3 Curing to Attain Specified Stripping or Release Strength

Usually stripping or release strengths are specified at a minimum of 2000 psi (13.8 MPa) for non-prestressed precast and 3000 psi (20.7 MPa) for prestressed units

C4.21 Accelerated Curing of Concrete

C4.21.1 General

Curing procedures should be determined on an experimental basis and are important, since they fix the total curing time within the 24-hr cycle essential to economic production.

Proper precautions should be taken with accelerated curing of units using surface retarders to make certain that the proper retarder is selected and that retardation is uniform and effective at time of stripping and treating the surface.

Standard	Commentary
ambient temperatures of the curing area.	

Standard

2. The beginning point for adding heat or elevating curing temperature over placed concrete temperatures shall be determined by using ASTM C403 to determine when the mortar strength has reached 500 psi (3.4 MPa) (initial set). For each mix design being considered, the results of three or more time of setting tests shall be plotted together. The best fitting smooth curve shall be drawn through the data points. These shall be used to determine the preset time which will be just beyond the development of 500 psi (3.4 MPa) mortar compressive strength.

   Care shall be taken to maintain the ASTM C403 test sample within the temperature range of 68 deg. to 77 deg. F (20 deg. to 25 deg. C). Once the preset time has been established for a particular mix design, it shall be adhered to. Heat application to the curing area may be required immediately in cold weather [below 50 deg. F (10 deg. C)] to maintain or regain placed concrete temperatures.

3. For accelerated curing, heat shall be applied at a controlled rate following the initial set of concrete in combination with an effective method of supplying or retaining moisture. Heat gain of not over 80 deg. F (27 deg. C) per hour is acceptable for the curing concrete as long as a proper delay period is used (ASTM C403).

4. The maximum sustained concrete temperature during the curing cycle shall be 160 deg. F (70 deg. C).

5. Maximum peak concrete temperature during the curing cycle shall be 190 deg. F (88 deg. C).

6. Maximum cooling rate from sustained accelerated curing temperature shall be 50 deg. F

Commentary

2. After placing, consolidation and finishing, the concrete should be allowed to attain its initial set before heat is applied; otherwise the elevated temperature may have a detrimental effect on the long-term strength of the concrete. Enough heat should be applied to the mold and its surrounding atmosphere to maintain the concrete at not less than 50 deg. F (10 deg. C). A period of 3 to 5 hr., depending on climatic conditions, type and brand of cement and admixtures, temperature of the mix and water-cement ratio is required for the concrete to attain its initial set. Delay periods in excess of 10 hr. may result in loss of effectiveness of heat curing. During the initial set period, provisions should be made to prevent surface drying.

3. If plant production procedures require early application of steam, the rate of temperature rise should be restricted to less than 20 deg. F (11 deg. C) per hr. Due to the slow rise of ambient temperature with radiant heat, application of the radiant heat cycle may be accelerated to meet climatic conditions.

4. The optimum curing temperature is from 140 deg. F to 160 deg. F (60 deg. C to 70 deg. C). In no case should the concrete temperature exceed 190 deg. F (88 deg. C) as higher temperatures may have a detrimental effect on strength. Duration of heat curing should be dictated by strength requirements for stripping as measured by test cylinders cured with the products. Caution should also be exercised when infra-red lamps are used. Localized "hot areas" should be avoided. When members produced with non-air entrained concrete are exposed to damp or wet conditions frequently or for lengthy periods, the maximum concrete temperature should be controlled to avoid possible problems with delayed ettringite formation and cracking of the concrete.

6. Units should be allowed to cool gradually to prevent thermal shock and rapid temperature changes, which may cause cracking. When the concrete is warmer than the air

Standard

per hr (28 deg. C per hr) and cooling at this rate shall continue until the concrete temperature is no more than 40 deg. F (22 deg. C) above the ambient temperature outside the enclosure in order to prevent surface crazing.

7. Self-recording thermometers shall be provided showing the time-temperature relationship through the curing period from placing concrete to stripping or transfer of prestress. At least one recording thermometer per contiguous form group with a common heat source shall be used to monitor the product at appropriate locations (usually the coolest position).

4.21.2 Curing with Live Steam

Steam curing shall be performed under a suitable enclosure to retain the live steam to minimize moisture and heat losses. The enclosure shall allow free circulation of the steam. Steam jets shall be positioned so they do not discharge directly onto the concrete, forms, or test cylinders.

4.21.3 Curing with Radiant Heat and Moisture

During the cycle of radiant heat curing, effective means shall be provided to prevent rapid loss of moisture in any part of the member.

4.22 Curing by Moisture Retention Without Supplemental Heat

4.22.1 General

For curing of the concrete without supplemental heat, the surface of the concrete shall be kept covered or moist until such time as the compressive strength of the concrete reaches the strength specified for detensioning or stripping.

Commentary

during cold weather, there is a tendency for soluble salts (efflorescence) to migrate to the surface immediately on stripping. In addition, cool water should not be used to accomplish aggregate exposure until the unit has cooled to ambient temperature.

7. To aid personnel who control the temperature during curing, it is recommended that the desired curing time-temperature relations be placed on the chart of the recording thermometer. With this information available, the desired temperature may be more easily maintained. Forms should be checked for uniformity of heat application. Temperature monitoring should be at the coolest position on a form.

C4.21.2 Curing with Live Steam

Monitoring techniques will require temperature checks at various points on a bed in different casts to effectively control curing.

C4.21.3 Curing with Radiant Heat and Moisture

Radiant heat may be applied to forms by means of pipes circulating steam, hot oil or hot water, by electric blankets, or heating elements on forms, or by circulating warm air under and around forms. Pipes blankets or other heating elements should not be in direct contact with concrete, forms or test cylinders.

Moisture may be applied by a cover of moist burlap, cotton matting, or other effective means. Moisture may be retained by covering the member with an impermeable sheet in combination with an insulating cover, or by applying a liquid seal coat or membrane curing compound.

Due to the slow rise of ambient temperatures with radiant heat, application of the heat cycle may be accelerated to meet climatic conditions. In all cases, the curing procedure to be used should be well established and carefully controlled to meet the requirements outlined in Article 4.21.

C4.22 Curing by Moisture Retention Without Supplemental Heat

C4.22.1 General

Acceptable methods of curing are:

1. Leaving the element in the form with the side forms in place and keeping the top surfaces continually moist by fogging, spraying or covering with moist burlap or cotton mats, or by covering top the top surface with impermeable covers or membrane curing compound.

2. Removing side forms and preventing loss of moisture from the sides and top surfaces by applicable methods described above.

Standard

4.22.2 Moisture Retention Enclosures

Any enclosures used for the purpose of retaining moisture during the curing period shall provide for retention of moisture to the extent that free water is at all times present on all concrete surfaces. Moisture retention enclosures shall be resistant to tearing and shall be positively fastened in place to avoid displacement by wind or other means during the curing cycle. Moisture retention enclosures shall remain in place until the completion of the curing cycle as described above.

4.22.3 Curing with Membrane Curing Compound

In the use of membrane curing compound to retain moisture within the concrete during curing the following shall be observed:

1. The coating of membrane curing compound shall cover the entire surface to be cured with a uniform film which will remain in place without gaps or omissions until full curing cycle is complete. Positive means shall be taken to detect and recoat areas of incomplete coating.

2. Membrane curing compound shall be applied at a rate of coverage which assures that the amount of compound applied per unit area of the surface being cured is at least the minimum coverage recommended by the manufacturer.

3. Membrane curing compound shall be applied to the top surface of the concrete element being cured as soon after casting as the surface water sheen disappears.

4. Membrane curing compound shall be compatible with coatings or other materials to be applied to the product in later construction stages.

Commentary

C4.22.3 Curing with Membrane Curing Compound

Curing compounds should only be used on the backs of units because of their tendency to discolor or stain. Some curing compounds function as bond breakers and may interfere with adhesion of repairs or surface coverings such as paints, fabrics, insulation materials or other specified protective coatings. Manufacturers should be consulted as to the effect that their compound has in these respects. In addition, membrane curing compounds should not be used in areas where joint sealants may be applied, unless they are entirely removed at the end of the curing period by sandblasting or by using an approved solvent, or unless conclusive tests show that the residue of the membrane does not reduce bond.

DIVISION 5 — REINFORCEMENT AND PRESTRESSING

Standard

5.1 Reinforcing Steel

5.1.1 General

Procedures for fabrication and placement of reinforcing steel shall be developed, implemented and understood by appropriate personnel. Procedures and practice shall be reviewed frequently to assure conformance.

5.1.2 Storage of Reinforcing Steel

The reinforcing steel deliveries shall be identified with a heat number which can be tied back to a mill certificate. Reinforcing steel shall be kept free of contamination, stored separately in a neat and orderly fashion and identified so that different types, grades, sizes and preheat requirements can be identified and recognized through the entire reinforcement preparation process. Bundles of reinforcing materials shall be kept straight and free of kinks until cut and bent to final shape to facilitate dimensional control within established placing tolerances. All reinforcing steel shall be stored on blocks, racks or sills, up off the ground. Special attention shall be given to prevent loose rust from forming or the steel from becoming contaminated with grease or oil or any material that would adversely affect the bond.

Equipment for handling coated welded wire or reinforcing bars shall have protected contact areas. Nylon slings or padded wire rope slings shall be used. Bundles of coated welded wire or reinforcing bars shall be lifted at multiple pick-up points to minimize abrasion from sags in the bundles. Hoisting with a spreader beam or similar device shall be used to prevent sags in bundles. Coated reinforcement shall not be dropped or dragged. Coated reinforcement shall be stored on timbers or other suitable protective cribbing with the dunnage spaced close enough to prevent sags in the bundles.

Epoxy-coated reinforcing bars shall be handled in a manner so that the protective coating is not damaged beyond what is permitted by ASTM A775/A775M or A934/A934M.

Each bundle of flat sheets of welded wire reinforcement shall have attached a suitable tag bearing the name of the manufacturer, style designation, width, length, and any other information specified by the purchase agreement. Steel strapping

Commentary

C5.1.2 Storage of Reinforcing Steel

Reinforcing steel should be subdivided into categories of preheat unless a uniform high preheat is chosen for all welded assemblies. To designate different preheat requirements, it is recommended that either a tagging or a coloring system be used to designate preheat requirements when the bars are received.

Good bond between reinforcement and concrete is essential if the steel is to perform its functions of resisting tension and of keeping cracks small. Therefore, the reinforcement should be free of materials injurious to bond, including loose rust. Mill scale that withstands hard wire brushing or a coating of tight rust is not detrimental to bond.

If it is necessary to store epoxy-coated bars outdoors for an extended period of time, usually more than 2 months, the bars should be protected from direct rays of sunlight and sheltered from the weather.

Standard

used to bundle the welded wire sheets shall not be used to lift the bundles.

5.1.3 Fabrication of Reinforcing Steel

The fabrication equipment shall be of a type, capacity, and accuracy capable of fabricating reinforcing cages to the required quality, including tolerances.

Review of fabrication shall be performed by quality control personnel to check that reinforcement has been cut and bent to correct shapes and dimensions, and is of correct size and grade.

Reinforcing bars shall be bent cold, unless otherwise permitted by the precast engineer, and shall not be bent or straightened in a manner that will injure the material. Bars with kinks or improper bends shall not be used. Diameter of bend measured on the inside of the bar shall be in accordance with ACI 318. Bars to be galvanized shall be bent in accordance with ASTM A767/A767M.

When zinc-coated (galvanized) reinforcment is damaged, the area to be repaired shall be coated with a zinc-rich paint (92 to 95 percent metallic zinc in the dry film) conforming to ASTM A767/A767M in accordance with ASTM A780.

Small spots of epoxy coating damage might occur during handling and fabrication and shall be repaired with patching material when the limits stated in the project specifications are exceeded. The maximum amount of repaired coating damage at the precast plant shall not exceed 2 percent of the total surface area per linear foot (0.3 m) of the coated bar. This means that a careful inspection and evaluation is needed prior to the approval of touch-up or re-coating.

Damaged epoxy-coating shall be repaired with patching material conforming to ASTM A775/A775M and in accordance with patching material manufacturer's recommendations.

When epoxy-coated reinforcing bars are welded or cut during fabrication, the weld area and the ends of the bars shall be coated with the same material used for repair of coating damage.

The starting and ending points of a welded wire reinforcement bend shall be located at least 4 wire diameters from the nearest welded cross wire. For deformed wires larger than W or D6, the diameter of the mandrel about which the bend is made shall be 4 wire diameters but only 2 wire diameters for all other wires.

Commentary

It is recommended that reinforcing bars be bent after galvanizing when possible. When galvanizing is performed before bending, some cracking and flaking of the galvanized coating at the bend is to be expected and is not a cause for rejection. The tendency for cracking of the galvanized coating increases with bar diameter and with severity and rate of bending.

Local removal of the galvanized coating in the area of welds, bends, or sheared ends will not significantly affect the protection offered by galvanizing, provided the exposed surface area is small compared to the adjacent surface area of galvanized steel. When the exposed area is excessive, and gaps are evident in the galvanized coating, the area should be repaired.

Standard

If reinforcing steel is fabricated by an outside supplier, that supplier shall furnish records of compliance to specification requirements and mill certificates for material used.

Cage assemblies, whether made for the entire casting or consisting of several sub-assemblies, shall be constructed so that they fit into molds without being forced. Cages shall have sufficient three dimensional stability so that it can be lifted from the jig and placed into the mold without permanent distortion. Also, the reinforcing cages shall be sufficiently rigid to prevent dislocation during consolidation in order to maintain the required cover over the reinforcement.

Reinforcing bars shall be tied using black annealed wire or welded in accordance with AWS D1.4 with the approval of the precast engineer. Zinc-coated reinforcement shall be tied with zinc-coated annealed tie wire, non-metallic coated tie wire, soft stainless steel or other acceptable materials. Epoxy-coated reinforcement shall be tied with plastic- or epoxy-coated tie wire or other acceptable material. All tie wires shall be bent back away from formed-surfaces to provide maximum concrete cover.

Where cages are tied, ends of ties shall not encroach on the concrete cover of the reinforcement. When cages are welded, care shall be taken that tack welding does not undercut reinforcing bars and thus diminish their area and strength.

All splicing of welded wire reinforcement or reinforcing bars, whether by lapping, mechanical connections or by welding shall be shown on the approved shop drawings. The concrete cover and bar spacings as a result of splicing shall conform to ACI 318. Mechanical connections or splices shall be installed in accordance with the splice device manufacturer's recommendations. After installation of mechanical connections on zinc- or epoxy-coated reinforcing bars, coating damage shall be repaired appropriately. All parts of mechanical connections used on coated bars, including steel splice sleeves, bolts, and nuts, shall be coated with the same material used for repair of coating damage.

5.1.4 Installation of Reinforcing Steel

The size, shape and spacing of all reinforcement shall be checked against the approved shop drawings. Variations in spacing of reinforcement exceeding allowable tolerances shall be corrected.

Commentary

C5.1.4 Installation of Reinforcing Steel

Standard

All reinforcement, at the time concrete is placed, shall be free of grease, form oil, wax, dirt, paint, loose rust or mill scale, or other contaminants that may reduce bond between steel and concrete or stain the surface of the concrete.

If there is more than one mat of reinforcement, bars shall be vertically aligned above each other in all horizontal directions during casting to minimize interference with placing and consolidating concrete.

Reinforcement shall be accurately located in the mold as indicated on the approved shop drawings and securely anchored to maintain its designed location within allowable tolerances while concrete is placed and consolidated. If spacers are used, they shall be of a type and of material which will not cause spalling of the concrete, rust marking or other deleterious effects. Metal chairs, with or without coating, shall not be used in a finished face. For smooth cast facing, chairs shall be plastic tipped or all-plastic to ensure absence of surface rust staining. If possible, reinforcement cages shall be securely suspended from the the back of the molds. Non-coated reinforcement shall be supported on reinforcement supports made of plastic.

Zinc-coated (galvanized) reinforcement supported from the mold shall rest on bar supports made of dielectric material or other acceptable materials. Galvanized reinforcement shall not be directly coupled to large areas of uncoated steel reinforcement unless plastic tie wire is used and local insulation is provided with dielectric materials, such as polyethylene or similar tape.

Epoxy-coated reinforcing bars supported from mold shall rest on bar supports made of plastic or other acceptable dielectric material. Reinforcing bars used as support bars for epoxy-coated material shall be coated with epoxy as well. Proprietary combination bar clips and spreaders used with epoxy-coated reinforcing bars shall be made of corrosion-resistant material or coated with non-conductive material.

Supports shall be sufficient in number and strength to carry the reinforcement they support and prevent displacement before and during concreting. They shall be spaced so that any sagging between supports will not intrude on the specified concrete cover.

After reinforcement is set to its final position within the mold, clearances or cover to the nearest deformation of surface for smooth wire shall be checked by measurement. Care shall be taken to maintain

Commentary

Reinforcement should be placed as symmetrically as possible about the panel's cross sectional centroid to minimize bowing and distortion of panels. Non-symmetrical placement may cause panel warpage due to restraint of drying shrinkage or temperature movements.

Spacers (bar supports) or chairs may mar the finished surface of an element which will eventually be exposed to weathering.

The bimetallic couple established by direct contact between galvanized steel and uncoated steel should not exhibit corrosive reactions as long as the depth to zinc/steel contact is not less than the cover required to protect uncoated steel alone under the same conditions or the galvanized mass is larger than the uncoated steel mass.

Protection of reinforcing steel from corrosion and the resultant possibility of surface staining is obtained by providing adequate cover. A protective iron oxide film forms on the surface of the bar as a result of the high alkalinity of the cement paste.

Standard

the critical dimensions determining the cover over reinforcement. The reinforcement type, sizes and spacing shall also be checked against the approved shop drawings. Variations in spacing of reinforcement exceeding allowable tolerances shall be corrected. The horizontal clear distance between reinforcement and mold shall be equal to the specified concrete cover or 1.5 times maximum aggregate size, whichever is larger.

For exposed aggregate surfaces, the concrete cover to surface of steel shall be measured from the mold surface and the typical depth of mortar removal between the pieces of coarse aggregate shall be subtracted. Attention shall also be given to scoring, false joints or rustications, and drips; the required minimum cover shall be measured from the thinnest location to the reinforcement.

Reinforcement shall be placed within the allowable tolerances, but the concrete cover shall be set so that the resulting concrete cover is never less than the specified cover.

Care shall be observed in placing of bars which extend out of the member and which are intended to provide structural connection for cast-in-place connections or use in sequential casting. The extensions shall be within ±1/2 in. (12 mm) of plan dimensions. Exposed reinforcing bars shall be protected from corrosion with a cold zinc coat or cement slurry to prevent rust washing down onto the exposed face during storage. Paste adhering to extended steel shall be removed to ensure bonding of bars to concrete in a later pour.

Reinforcement shall not be bent, without approval of the precast engineer, after being embedded in hardened concrete.

Commentary

As long as this alkalinity is maintained, the film is effective in preventing corrosion. The protective high alkalinity of the cement paste is usually lost only by leaching or carbonation. Therefore, concrete of sufficiently low permeability with the required cover over the steel will provide adequate protection. Low permeability is characteristic of well-consolidated concrete with a low water-cement ratio and high cement content. This composition is typical of architectural precast concrete.

For concrete surfaces exposed to weather, prestressing strand should be protected by a concrete cover of not less than 1 in. (25 mm). Reinforcing steel should be protected by concrete cover equal to nominal diameter of bars but not less than 3/4 in. (19 mm).

For concrete surfaces not exposed to weather, ground, or water, prestressing strand should be protected by a concrete cover of not less than 3/4 in. (19 mm). Reinforcing steel should be protected by concrete cover of not less than the nominal diameter of bars but not less than 5/8 in. (16 mm).

Cover requirements over reinforcement should be increased to 1-1/2 in. (38 mm) for non-galvanized reinforcement or be 3/4 in. (19 mm) with galvanized or epoxy coated reinforcement when the precast concrete elements are acid treated or exposed to a corrosive environment or to severe exposure conditions. In addition, the 3/4 in. (19 mm) cover is realistic only if the maximum aggregate size does not exceed 1/2 in. (12 mm) and the reinforcing cage is not complex.

Reinforcing bar sizes No. 3 through No. 5 may be bent cold the first time provided reinforcing bar temperature is above 32 deg. F (0 deg. C). For bar sizes larger than No. 5 reinforcing bars should be preheated before bending. Heat may be applied by any method which does not harm the reinforcing bar material or cause damage to the concrete. A length of reinforcing bar equal to at least 5 bar diameters in each direction from the center of the bend should be preheated but preheating should not extend below the surface of the concrete. The temperature of the reinforcing bar at the concrete interface should not exceed 500 deg. F. (260 deg. C). The preheat temperature of the reinforcing bar should be between 1100 to 1200 deg. F (593 to 649 deg. C).

Standard

If reinforcing steel or hardware anchors cannot be located as shown on the drawings, all changes shall be reviewed and approved by the precast engineer and drawings in turn corrected to show the as-cast position. Under no circumstances shall main reinforcement or prestressing steel be eliminated to accommodate hardware. Reinforcement close to anchors, on which hardware units rely for anchorage, shall also be provided and maintained.

5.2 Tensioning

5.2.1 General Tensioning Requirements*

In all methods of tensioning, stressing of tendons shall be accomplished within close limits, as the force is critical for both performance and structural safety of the member and the structure of which it forms a part. Because the force cannot be checked accurately later in the production process, the stressing operation is important and shall be subject to careful production and quality controls.

5.2.2 Tensioning of Tendons

In all methods of tensioning, force in the tendons shall be determined by monitoring applied force and independently by measuring elongation. Applied force may be monitored by direct measurement using a pressure gauge piped into the hydraulic pump and jack system, dynamometer, load cell, or other accurate devices. At the completion of stressing operations, the two control measurements, force and elongation, shall agree with their computed theoretical values within a tolerance of ±5 percent. If discrepancies are in excess of 5 percent, the tensioning operation shall be suspended and the source of error determined and evaluated by qualified personnel before proceeding. Additionally, the control measurements of force and elongation shall algebraically agree with each other within 5 percentage points. If the measurements do not agree within 5 percent, a load cell

---

* The provisions set forth in this section refer to the application and measurement of stresses to prestressed concrete members manufactured by the process of pretensioning, post-tensioning, or a combination of the two methods. For conditions not covered here, such as draping or bundling of prestressing steel, or debonding of strand, and production of dry-mix, machine-cast products, refer to PCI MNL-116.

Commentary

The preheat temperature should be maintained until bending or straightening is complete. The preheat temperature should be measured by temperature measurement crayons, contact pyrometer, or other acceptable method. The heated bars should not be artificially cooled (with water or forced air) until after cooling to at least 600 deg. F (316 deg. C).

C5.2 Tensioning

C5.2.1 General Tensioning Requirements

The ultimate capacity of a prestressed concrete member is usually not affected by moderate variations in the stressing of individual tendons; however, inaccurate and variable stressing can result in differential camber, inaccurate concrete stresses, lateral bowing of members, and reduction of the cracking load, all of which can contribute to the manufacture of unsatisfactory members.

C5.2.2 Tensioning of Tendons

In tensioning of tendons, the gauging system indicates the proper force has been applied. A check of elongation indicates the correct size of tendon has been used and operational losses are within tolerance limits, and it provides a check on the gauging system. Elongation aids in confirming the physical properties and characteristics of the strand used. Information on elongation corrections is in Article 5.3.10 and information on force corrections is in Article 5.3.11.

Standard

may be added at the "dead end" and if force measurements agree within 5 percent between the gauge at the live and the load cell at the dead end, the elongation agreement can be waived.

After an initial force has been applied to the tendon, reference points for measuring elongation due to additional tensioning forces shall be established.

Calculations for elongation and gauge readings shall include appropriate allowances for friction, chuck seating, movement of abutments, bed shortening if under load, thermal corrections, and any other compensation for the setup.

5.2.3 Methods of Force Measurement

Methods of measurement of the tensioning force shall consist of one or more of the following:

1. Pressure gauges to measure force from the pressure applied to hydraulic jacks.

2. Dynamometers connected in tension into the stressing system.

3. Load cells connected into the stressing system so the action of the stressing operation imparts a compressive force to the sensing element.

4. Digital readouts connected to a pressure transducer to measure force from the pressure applied to the hydraulic jack.

5. Force computed from the actual elongation of the strand based on its physical properties and compensation adjustment. This method is used as a review of the four methods above.

Commentary

C5.2.3 Methods of Force Measurement

1. Pressure gauges or transducers should have dials or digital readout calibrated to show jacking force by means of an approved and calibrated load cell. The gauges can show hydraulic pressure which, through correlation with the area of the ram in use, determines the actual load used for the tensioning process.

    It is preferable to calibrate rams and gauges together as a system. However, gauges may be calibrated against a master gauge of known accuracy, provided the rams are calibrated against the same master gauge.

2. Dynamometers can be used for initial tensioning operations due to the lower level of forces involved in initial tensioning.

3. Properly calibrated load cells will provide the most accurate measure of tendon stress at their point of application. Jacking systems are available with load cells in the jack head or pressure transducer and a digital readout instead of a gauge.

5. To determine the stressing force (P) from elongation (Δ), use the equation:

$$P = \frac{\Delta AE}{L}$$

where A is cross-sectional area of the strand and E is modulus of elasticity of the strand and L is the stressing length chuck-to-chuck.

Sample calculations for tensioning setups are shown in Appendix H.

Standard

5.2.4 Gauging Systems

Hydraulic gauges, dynamometers, load cells or other devices for measuring the stressing load shall be graduated so they can be read within a tolerance of ±2 percent. Gauges, jacks, pumps, hoses and connections shall be calibrated as a system in the same manner they are used in tensioning operations. Calibrations shall be performed by an approved testing laboratory, calibration service, or under the supervision of a registered professional engineer on the staff of a production plant or as a consultant, according to the equipment manufacturer's recommendations. A certified calibration curve shall accompany each tensioning system. Pressure readings can be used directly if the calibration determines a reading is within a ±2 percent tolerance of actual load. Calibrations shall be performed at any time a tensioning system indicates erratic results, and at intervals not greater than 12 months.

Pressure gauges, pressure transducers for hydraulic systems, or other measuring devices, such as digital readout, shall have a full range of measurement of 1-1/2 to 2 times their normal working pressure, whether for initial or final force. If the same unit is used for both initial and final tensioning, the jacking system shall have separate gauges or separate scales to ensure accurate measurement of both the initial and final force. Gauge/transducer readings based on system pressure shall not be made below 10 percent or above 90 percent of the full-scale capacity of the gauge/transducer unless the gauge/transducer is calibrated within that range and 2 percent accuracy verified.

Tensioning methods employing hydraulic gauges shall have appropriate bypass valve snubbers and fittings so that the gauge pointer will not fluctuate but will remain steady until the jacking load is released.

5.2.5 Control of Jacking Force

Pressure bypass valves may be used for stopping the jack at the required force or for manually stopping the application of force with the valve. The accuracy of setting of automatic cutoff valves shall be verified by running to the desired cutoff force whenever there is reason to suspect improper results, and at a minimum, at the beginning of the operation each day.

5.2.6 Wire Failure in Strand or Tendon

Failure of individual wires in a pretensioning strand or post-tensioning tendon is acceptable provided

Commentary

C5.2.4 Gauging Systems

Gauges or digital readouts for single strand jacks may be calibrated by means of an approved and calibrated load cell. Gauges for large multiple strand jacks acting singly or in multiple should be calibrated by proving rings or by load cells placed on either side of the movable end carriage.

In multiple strand tensioning, use of a master gauge system to monitor accuracy of hydraulic gauges is acceptable as an ongoing calibration method, since the cycles of tensioning are only a fraction of the cycles in a single strand system.

Forces to be measured should not be less than one-fourth and not more than three-quarters of the total graduated capacity, unless calibration information clearly establishes consistent accuracy over a wider range. Gauges should have indicating dials at least 6 in. (150 mm) in diameter. Gauges should also be mounted at or near working eye level and within 6 ft (1.8 m) of the operator, positioned so that readings may be obtained without parallax. A gauging system should have at least two gauges, a working gauge and a check gauge.

C5.2.5 Control of Jacking Force

When manual cutoffs are used for control of jacking force, rate of force application should be slow enough to permit the operator to stop the jack within the limits of specified force tolerances.

C5.2.6 Wire Failure in Strand or Tendon

The 2 percent limit of the total area of tendons in a member represents a relatively limited number of broken wires that can be accepted in a setup. When the 2 percent tolerance is ex-

Standard

the 5 percent allowable variation in prestress force for all strands in the entire element is not exceeded and provided the total area of broken wires is not more than 2 percent of the total area of tendons in a member, and providing the breakage is not symptomatic of a more extensive distress condition.

The entire strand shall be considered ineffective if a wire breaks in a 3-wire strand.

Welding shall not be performed near any prestressing strand. The prestressing strand shall not be exposed to spatter, direct heat, or short-circuited current flow.

5.2.7 Calibration Records for Jacking Equipment

Calibration records shall show the following data:

1. Date of calibration.
2. Agency, laboratory or registered engineer supervising the calibration.
3. Method of calibration; i.e., proving ring, load cell, testing machine, etc., and its calibration reference.
4. The full range of calibration with gauge readings indicated against actual load (force).

Calibration records for all tensioning systems in use shall be on hand for use in preparing theoretical tensioning values and shall be maintained until the next calibration. Personnel involved in preparing tensioning calculations shall have a copy of these records for reference.

5.3 Pretensioning

5.3.1 Storage of Prestressing Steel

Prestressing steel reels and coiled tendons shall be stored with identifying tags listing the heat number to relate the reel or tendon to a mill certificate. It is recommended that the reel or coil numbers be identified on the stressing sheets. Care shall be taken in storage to avoid confusion between different types [low relaxation or stress relieved (normal relaxation) strand] or diameters. Material handling of prestressing steel in the plant shall be done carefully to avoid abrading, nicking or kinking the strands, bars or wire as they are moved through the plant or set up for stressing. Special attention shall be given to protecting sheathing when unloading and storing coiled, sheathed tendons.

Care shall be taken in the storage of prestressing steel to prevent corrosion due to humidity, galvanic

Commentary

ceeded, an adjustment is required which may sometimes necessitate detensioning and replacing the affected strands.

In any event, the setup with broken wires should be examined to determine the reason for the break, and engineering should be alerted for an evaluation. Strands which have welds within the coils sometimes demonstrate broken wires at these weld points. Other problem areas include incorrectly aligned strand chucks which create a bind at the front of the chuck during seating, or improperly maintained chucks.

Weld spatter can cause stress concentrations, and the temperature rise due to the direct effect of welding heat or the indirect effect of current flow through the high-tensile prestressing steel can cause a sudden loss of tension (strand rupture).

C5.2.7 Calibration Records for Jacking Equipment

Tensioning systems must be calibrated on a regular basis or whenever erratic results are encountered to provide effective control of tensioning. Tensioning calculations should be done with correct and current calibration data, therefore these data should be distributed to personnel preparing these calculations.

C5.3 Pretensioning

C5.3.1 Storage of Prestressing Steel

The minimum yield of low relaxation and stress-relieved (normal relaxation) strand are different so the strands need to be identified to avoid overstressing the stress-relieved strand to low relaxation strand loads.

High strength steel is much more susceptible to corrosion than steel of lower strengths. Where prestressing steel is exposed to

Standard

or battery action which can occur when dissimilar metals are adjacent to an ionized medium common to both, or electric ground currents.

Bands on strand packs shall not be cut by torch. Cutting or welding around strands in storage or prior to casting shall not be allowed.

5.3.2 General

In all methods of pretensioning, the force shall be applied in two increments. An initial force shall be applied to the individual strands to straighten them, eliminate slack, and provide a starting or reference point for measuring elongation due to additional tensioning forces. The final force shall then be applied and elongation of strands computed and measured from the reference points. This method of operation shall be mandatory except as provided in Article 5.2.2.

Each plant shall develop written stressing procedures providing step by step instruction to personnel performing the stressing operation. Personnel shall be well trained, and personnel authorized to perform and/or record the stressing process shall be identified as part of the written procedures.

Stressing procedures shall include instructions for:

1. Operation and control of jacking equipment.
2. Operation and control of gauging system.
3. Stressing to an initial force and marking strand in preparation for measuring elongation.
4. Stressing to a given final force, measuring and recording the corresponding elongation.
5. Checking for strand anchor seating.
6. Procedures in case of out-of-tolerance results.
7. Procedures in case of wire failure.
8. Alternative stressing methods or measurements.
9. Detensioning and stripping.

5.3.3 Strand Surfaces

Special care shall be exercised to prevent contamination of strands from form release agent, mud, grease or other contaminants that would reduce the bond between steel and concrete. Form release agent shall be applied to the mold so that it does not contaminate any strand.

Commentary

wet weather or excessively humid conditions in storage, corrosion damage may occur within a few weeks. Storage under cover is preferred as a means of minimizing corrosion. Corrosion which deeply etches or pits the surfaces cannot be tolerated on prestressing steels; however, a light coating of tight surface rust is beneficial to bond.

Strand properties are altered by concentrated heat or arcing electrical current. This alteration can result in lowered ultimate strength which could result in strand failure under load.

C5.3.2 General

Two methods of pretensioning are in general use. Single strand tensioning consists of stressing each strand individually. Multiple strand tensioning consists of tensioning several strands simultaneously. Either method may be used subject to proper allowances and controls. Both methods should utilize initial tension, then a final tension increment. Usually, prestressing in a member, after all losses, is in the range of 150 to 800 psi (1 to 5.5 MPa).

C5.3.3 Strand Surfaces

Prior to stringing of strands, bottom forms should be inspected for cleanliness and accuracy of alignment due to the difficulty in making corrections after the strands have been tensioned. The entire force of a strand is transferred into the concrete through bonding of the hardened concrete to the strand. Therefore, it is extremely important that the strands be clean prior to

Standard

5.3.4 Stringing of Strands

To avoid possible entanglement of strands and minimize unbalanced loads on the molds during tensioning, a planned sequence of stringing and tensioning shall be followed.

Strands shall be pulled from the correct side of the pack, as identified by the manufacturer. Each length of strand shall be cut between the strand chuck and the coil or reel. Stringing of lengths of strand incorporating points previously gripped by strand chucks within lengths to be stressed shall not be permitted.

After stringing and tensioning, the strand shall be inspected for contamination by form release agent or other surface coatings and, if contaminated, shall be cleaned using an approved method.

5.3.5 Strand Chucks and Splice Chucks

Strand chucks and splice chucks shall be capable of securely anchoring maximum tensioning forces. Chucks shall be used as complete units. Strand chucks designed with spring-equipped caps shall be used with caps. Strand chuck components shall be inspected between each use, cleaned, and lubricated as necessary. Barrels, wedges or caps that become visibly worn, cracked, distorted, or which allow slippage, shall be discarded. Strand chucks shall be assembled of compatible components from the same manufacturer to avoid improper fit and seating on strands. Inspection and maintenance of strand chucks in use shall include matching of chuck barrels and wedges by strand size and manufacturer. During inspection and reassembly, care shall be taken to avoid assembling improper chucks; i.e., 1/2 in. (13 mm) barrels with 7/16 in.

Commentary

concrete placement to ensure that bonding takes place. Since it is extremely difficult to effectively clean strands that have been contaminated, it is desirable to plan a program of prevention of contamination rather than depending upon cleaning after contamination occurs.

Release agents which do not dry but remain as an oil should be evenly applied to assure release, but without excess or puddles which would contaminate strand placed in the mold. Any excess release agent should be removed from the mold surface.

C5.3.4 Stringing of Strands

An orderly procedure of stringing and tensioning strands facilitates the keeping of records and is essential when data from a force recording device are to be identified with a particular strand.

Strand may be furnished either in coils, reel-less packs, or on reels. Strands may be strung singly or in multiples. Strand paid-out from a coil or reel-less pack will rotate each time a revolution is pulled from the coil. Provision should be made to relieve these rotations.

Strands may be threaded through bulkheads or cages of reinforcing steel. In this case, care should be taken so the strand passes through freely and binding does not occur during the stressing operation.

The practice of continuous stranding is not allowed due to the potential of placing nicked strand within a member. All strand chucks notch or nick the wires of the strand. The nicks result in local stress concentrations which may result in failure of the strand during stressing and which through fatigue lower the ultimate strength of the strand.

C5.3.5 Strand Chucks and Splice Chucks

Strand chucks generally consist of a barrel, grooved wedges with an O-ring pulling them together, and a spring-equipped cap.

Proper care of strand chucks cannot be overemphasized. In Appendix D guidelines are given for daily inspections of strand chucks. The beginning cracks which can be observed in wedges and barrels are evidence that the elements should be taken from service immediately to avoid a potential failure.

Strand chucks which are not equipped with spring caps have a tendency to seat the wedges unevenly, producing stress-risers on sides of the strand at the forward contact point. When these chucks are used, proper attention should be given to assure even seating of the sections.

Standard

(11 mm) wedges, etc. Chuck maintenance is an important aspect of tensioning and is further outlined in Appendix D.

5.3.6 Strand Splices

Strand lengths spliced together shall have the same lay of wire to avoid unraveling. Only one splice per strand shall be permitted. The ends of the strand to be spliced shall be cut by shears or abrasive saws or grinders.

The location of strand splices shall be such that they will not fall within a member unless they are of a type that will develop the full ultimate strength of the strand.

For single strand tensioning, as the seating losses of individual splices can be checked and corrected for differential seating, the number of strands per bed that may be spliced is not restricted.

For multiple strand tensioning either all or none of the strands may be spliced.

5.3.7 Strand Position

Strands shall be positioned in accordance with detailed dimensions shown in the production drawings. Strands shall be supported as required to maintain the vertical and horizontal position within the tolerances as specified in Article 7.1.2.

5.3.8 Initial Tensioning

Care shall be taken to ensure that a valid starting point is established for elongation measurement by initial tension. After strands have been positioned, an initial force in the range between 5 and 25 percent of the final force shall be applied to each strand which will have elongation measured to pro-

Commentary

C5.3.6 Strand Splices

When strand sections are spliced together with opposing lay (twist), the splice will rotate, resulting in considerable stress loss in the strand.

As the structural properties of the strand in the immediate vicinity of torch-cutting are affected by heat, it is required that the ends of the strand to be spliced be cut by shears or abrasive saws or grinders.

C5.3.7 Strand Position

The importance of the correct quantity and position of prestressing strand cannot be overemphasized. These factors are critical to the product performing as designed. At the very least, strand position should be checked initially at the ends of the members and at all intermediate bulkheads along the form. Member behavior is relatively insensitive to horizontal location of tendons in typical flat panels or members significantly wider than thick.

Check beds and equipment (headers, etc.) initially to assure desired strand position. It is particularly important to check strand position requirements when incidental reinforcing steel is supported partially or entirely by the strands, as the weight of the incidental steel will pull the strand out of position if not monitored closely. On long-line products or heavily reinforced cage sections, strand should be chaired to assure design position. On panels the vertical position of prestressing steel is critical and should be closely controlled and monitored. The sequence of placement and location of reinforcing steel, inserts, and blockouts should be carefully planned to avoid interference with the designed vertical position of the prestressing steel.

C5.3.8 Initial Tensioning

Initial tension is an important point to establish during the tensioning operations. For that reason, the initial tension should be enough to remove the slack from the strand so that a point can be established for elongation measurement that is valid, but not too excessive as to eliminate too much of the elongation which will take place to reach final load.

Standard

vide an initial force in all strands that will result in a uniform final force. Regardless of the method used, initial force shall be measured within a tolerance of ±100 lb (±45.4 kg) per strand if initial force is equal to or less than 10 percent of final force, or ±200 lb (±90.7 kg) per strand if initial force is greater than 10 percent of final force. Elongation measurements as a measure of initial force are impractical and shall not be used. In self-stressing forms, care must be taken to tension strand symmetrically on the form to prevent warping of the form due to eccentric loading.

5.3.9 Measurement of Elongation

At the completion of initial tension, reference marks shall be established from which elongation by final tensioning forces can be measured. Elongations are then accurately measured from these reference points. Elongations shall be measured as required in Articles 5.2.2 and 5.3.12.

5.3.10 Elongation Calculation and Corrections

Elongation measurement shall take into account all operational losses and compensations in the tensioning system. For computation of the elongation, the modulus of elasticity and steel area of the tendon shall be determined from mill certificates provided by the manufacturer of the strand. An average area and modulus may be used provided the force indicated falls within the tolerance limit specified herein. Corrections to the basic computed elongation for a bed setup vary between casting beds and shall be evaluated and compensated for in computing elongation. Principal operating variables are:

1. Chuck seating.

    a. Dead end seating.

Commentary

A sufficient initial force should be applied to produce the majority of strand seating into the dead end chuck. If a final stressing force of less than $0.40f_{pu}$ is used, special procedures are required to prevent strand slippage in the chuck. Consult the strand supplier for details.

C5.3.9 Measurement of Elongation

For single strand tensioning, elongation measurements should be made by marking the strand to allow measurement of that strand.

For multiple strand setups, elongation can be measured by travel of stressing header. Both types should be monitored to confirm operational losses and adjustments. The operational corrections considered in the measurement of elongation are outlined in Article 5.3.10 and should be included in each setup.

Measurement for elongation must be actual measurements rather than estimates so that an evaluation can be made in determining the validity of corrections and the validity of the tensioning process. The degree of accuracy necessary in measuring elongations depends on the magnitude of elongation, which is a function of the length of the bed. Accuracy of measurement to closer than the nearest 1/4 in. (6 mm), corresponding to a maximum error of 1/8 in. (3 mm), is considered impractical. For forces used normally with 1/2 in. (13 mm) diameter high tensile strength strand and wire, accuracy of 1/4 in. (6 mm) corresponds to force variations of approximately 3 percent in a 50 ft (15 m) length of bed or member and 1 percent for a 150 ft (46 m) bed or member. Accuracies of this order can and should be attained.

C5.3.10 Elongation Calculation and Corrections

Sample calculations for tensioning setups are shown in Appendix H.

The computation of elongation is based on force applied to the strand which is a percentage of the ultimate capacity of the strand. Forms should be surveyed and monitored to establish appropriate elongation corrections. Any alterations of the form will require reevaluation of corrections and many new setups will require variation in assumptions. The following are the most common corrections for consideration:

1. **Chuck seating**

    a. Dead end seating. The initial force placed on the

Standard	Commentary
	strand seats the strand into the dead end chuck; however, from that initial tension force to the final tension force, additional seating will typically occur and will show up as added elongation in the measurement.
b. Live end seating.	b. Live end seating. As a strand is seated into the live end chuck, seating loss occurs. In multiple strand tensioning, this is a relatively nominal amount similar to dead end seating loss. However, in single strand tensioning, stress is transferred from the jack to the chuck after the final load is reached. If the chucks are in good condition with properly functioning spring caps, this seating loss is not too great, but it must be anticipated and will generally range from 3/8 in. to 3/4 in. (10 to 19 mm). This should be monitored by plant personnel to determine an appropriate value for the tensioning system in use at the plant.
c. Splice chuck seating.	c. Splice chuck seating. Similar to dead end seating loss, the strands are seated in the splice chuck by the initial tension loads. A small amount of seating loss will still take place from the initial force to the final force. Strands should be marked on each side of the splice to confirm the assumed seating loss. This should be done often enough to confirm assumptions of seating used in calculations.
2. Form shortening (self-stressing forms).	2. **Form shortening.** Self-stressing forms will shorten under the stressing load on its ends. Since elongation is typically measured as a reference to the end of the form, the form shortening shows up as increased elongation. As one single strand is tensioned, the form shortening is very slight. However, members do not move instantaneously under load, and the form is no exception. Many times the form has enough friction on the subsurface to restrain movement until a given point. That is why an average value is best used, since there is no way to predict when shortening will take place.
3. Abutment rotation or movement of anchorages (fixed abutment forms) – elongation of abutment anchoring rods.	3. **Movement of anchorages.** This is most commonly a phenomenon for abutment forms where the loads are imposed onto the abutments which deflect elastically. The foundation has potential movement as well. The movements are usually very small for well constructed abutments, but they should be evaluated and compensated for in calculations.
4. Thermal effects.	4. **Thermal effects.** In the event that thermal changes are anticipated, as the strand is warmed from a cold morning stressing to concrete placement, then the calculated elongation must be changed to correspond to the changes in stress.
5. Gauge correction based on calibration data.	

Standard

5.3.11 Force Corrections

Operational conditions resulting in variations of stress as indicated by jacking pressure consist of:

1. **Friction in jacking system.** Rams used in jacks for single strand tensioning are typically small; therefore, friction losses in the jacking system can usually be ignored provided gauges and systems are calibrated in accordance with Article 5.2.7. The anchorages are not part of the jacking system.

   Multiple strand tensioning methods, because of the large jacking ram and the heavy sliding or rolling anchorage to which all strands are attached, are subject to friction which must be overcome by the jack in addition to the force required to tension the strand. There may be substantial variations in friction between various strand patterns. To minimize friction, the sliding surfaces shall be clean and well lubricated. If two methods of force measurement exceed the allowable variation, a third method shall be used.

2. **Thermal effects.** For abutment anchorage set ups, where strands are anchored to abutments that are independent from the form, thermal adjustments are required if the temperature of the steel at the time it is tensioned and at the time the concrete starts to set differ by more than 25 deg. F (15 deg. C) and the net effect is greater than a 2-1/2 percent force differential. Consideration shall be given to partial bed length usage and adjustments made when the net effect on the length of bed used exceeds the allowable. The thermal coefficient of expansion of steel shall be taken as 6.5×10^{-6}/deg. F (12×10^{-6}/deg. C). Since tensioned strands are held at a fixed length, variation between ambient temperature at time of stressing and concrete temperature at time of placing results in changes of stress. Lowering of strand temperature increases force while a temperature rise results in force loss. For strands stressed to approximately 70 to 75 percent of the strand ultimate tensile strength, a temperature variation in the strand of 10 deg. F (5 deg. C) will result in a variation of 1 percent. Allowance shall be made in the stressing for temperature variation of 25 deg. F (15 deg. C) or more by understressing or overstressing at the rate of 1 percent for 10 deg. F (5 deg. C). of anticipated temperature variation, depending respectively on whether a reduction or rise of temperature is anticipated. This adjustment is typically not required for self-

Commentary

C5.3.11 Force Corrections

Any instance where friction is excessive for multiple strand systems, rendering the gauge pressure ineffective for a control, load cells would be required for the substantiation as a third method for verification of the stressing force.

2. **Thermal effects.** This item is important for abutment beds as the abutments are not affected by the temperature rise of the strand from the temperature at stressing on a cold morning to the concrete temperature at placement in the forms at mid-day or afternoon. The actions would be reversed for strands tensioned at elevated temperatures with cooler concrete cast around it. Self-stressing beds are not affected by this phenomenon because the bed itself undergoes a similar change. As the strand is warmed and expands, the bed does as well. Since the strand is anchored to the bed, the force in the strand is relative to the bed length. As the strand is expanding and trying to relax, the bed is expanding and holding the strand at its desired value.

Standard

stressing beds.

3. **Live end seating overpull.** Release of strands from single strand jacks results in a force loss as the strand seats into the live end chuck. To compensate for this loss, strands shall be stressed and elongated additionally to offset the seating. The extra force shall be added to the desired theoretical force.

5.3.12 Final Stressing of Strands

For single strand tensioning, after application of the initial force and establishment of reference marks for measuring elongation, the full strand force shall be applied. Strand force indicated by gauging systems shall control the tensioning, with elongation checked on every strand. An exception is the case of a completely open bed with no bulkheads or other possible sources of friction. In such instances, strand elongation shall be checked on the first and last strands stressed and at least 10 percent of the remaining strands.

For multiple strand tensioning, following application of initial force and seating of each strand on the anchorage header, reference marks shall be established for measuring elongation and seating. Reference marks for seating shall be made by marking a straight line across the strands in each row along the face of the anchorage. For uniform application of force to strands, the face of anchorage at final load shall be in a plane parallel to its position under initial force. Parallel movement shall be verified by measurement of movement on opposite sides of the anchorage and a check of its plumb position before and after application of the final force.

The allowable jacking force during tensioning shall not exceed the values stated in the latest edition of the PCI Design Handbook.

5.3.13 Detensioning

Stress shall not be transferred to pretensioned members until concrete strength, as indicated by test cylinders or other properly calibrated nondestructive test technique, is in accordance with specified transfer strength.

If concrete has been heat-cured, detensioning shall be performed immediately following the curing period while the concrete is still warm and moist.

Commentary

3. **Live end seating overpull.** To add additional elongation to offset seating losses requires additional force to stretch the strand. The extra force must be added to the desired theoretical force.

Since elongation is measured by travel of the anchorage, a reference mark is usually made at the face of the anchorage on each side of the bed.

C5.3.12 Final Stressing of Strands

Straight strands are the most straightforward tensioning setup. Care is required to be sure a valid starting point is established by pulling to initial force for elongation measurement and that all appropriate corrections are made in the setup calculations.

C5.3.13 Detensioning

Tests have shown bond transfer length for wet mix concrete is not appreciably affected by concrete strengths in the range of 2500 psi to 4000 psi (17.2 to 27.6 MPa) at release. Concrete strength does influence camber and dimensional changes due to strains in the concrete. Minimum concrete transfer strength of 3,000 psi (20.7 MPa) is recommended.

If concrete is allowed to dry and cool prior to detensioning, dimensional changes take place which may cause contraction cracking or undesirable stress building up in the concrete. Before detensioning, temperature differentials should be minimized between the concrete mass and the ambient air to which the top

Standard

In single strand detensioning, both ends of the bed shall be released simultaneously and symmetrically to minimize sliding of members. Forms, ties, inserts, or other devices that would restrict longitudinal movement of the members along the bed shall be removed or loosened. Alternately, detensioning shall be performed in such manner and sequence that longitudinal movement is precluded.

In multiple strand detensioning, strands are released simultaneously by hydraulic dejacking. The total force is taken from the header by the jack, then released gradually. The overstress required to loosen lock nuts or other anchoring devices at the header shall not exceed the force in the strand by more than 5 percent.

The sequence used for detensioning strands shall be according to a pattern and schedule that keeps the stresses nearly symmetrical about the vertical axis of the members, and shall be applied in a manner that will minimize sudden shock or loading. Maximum eccentricity about the vertical axis of the member in the casting bed shall be limited to one strand or 10 percent of the strand group. Limitation of vertical axis eccentricity shall be at initial cutting of ends of bed and as strands are cut between members in setup. For unusual shapes and heavily stressed shapes, production drawings shall show detensioning procedures.

5.3.14 Protection of Strand Ends and Anchorages

Special attention shall be paid to finishing of ends of members in the area of strand ends and anchorages, as specified on the shop drawings before members are yarded. Unless such areas are maintained in a permanently dry condition after erection, strand ends and anchorages shall be protected against moisture penetration.

Commentary

surface is exposed. Strands should be detensioned immediately upon stripping off covers, or a way developed to detension strands before stripping the covers (or partially detensioned before stripping covers). The use of self-stressing forms reduces the effect of dimensional changes.

For stress to be released gradually, strands should not be cut quickly but should be heated until the metal gradually loses its strength. This becomes much more significant as the ratio of prestressing force to the area of member increases.

Loss of prestress due to slippage between concrete and strand also reduces load carrying capacity since moment capacity reduces as slippage increases into the section. The affected length of slipped strand is determined by considering an overlap of the zone of flexural bond with the zone of strand stress development bond.

C5.3.14 Protection of Strand Ends and Anchorages

Lack of proper protection of end anchorages allows an access point for moisture and a developing point for corrosion. If salts are present in the moisture or water that accesses an anchorage, then corrosion is enhanced and often accelerated, making adequate end protection mandatory.

When exposed to view, anchorages (stressing pockets) should be recessed and packed with a minimum of a 1-in. (25 mm) thickness of non-metallic, non-shrink mortar and receive a sack finish. Prior to installing the pocket mortar, the inside concrete surfaces of the pocket should be coated or sprayed with an epoxy bonding agent. This mortar seal should be adequately covered for curing since shrinkage or contraction cracks will permit moisture penetration. When not exposed to view, strand ends should be coated with a rust inhibitor, such as bitumastic, zinc rich- or epoxy-paint to avoid corrosion and possible rust spots.

Standard

5.4 Post-Tensioning of Plant-Produced Products

5.4.1 General

The stressing of post-tensioned members is governed by many of the considerations applicable to pretensioned concrete.

Stress in the tendons shall always be measured by gauge readings and verified by elongation. Due to frictional losses peculiar to post-tensioned members and generally due to their relatively short length (as compared to most pretensioning beds) the predetermination of jacking loads and elongations and accuracy and reconciliation in measurement are particularly important. The elastic shortening of the concrete member during tensioning shall be given due consideration in computing apparent elongations.

Records shall be maintained for plant post-tensioning operations in a similar fashion to other plant operations.

Post-tensioning systems shall be installed in accordance with manufacturers' directions and proven procedures. Manufacturers' recommendations shall be observed regarding end block details and special reinforcement in anchorage zones applicable to their particular systems.

Plastic coated unbonded tendons with a low coefficient of angular friction looped within the panel, and anchorages installed at one end only or at both ends may be used. Curvature in the tendon profile shall preferably not be closer than 3 feet (0.9 m) from the stressing anchorage. Tendons shall be firmly supported at intervals not exceeding 4 feet (1.2 m) to prevent displacement during concrete placement.

Commentary

C5.4 Post-Tensioning of Plant-Produced Products

C5.4.1 General

Many architectural panels which do not lend themselves to being pretensioned because of difficulties with long line casting, such as jacking bulkhead or self-stressing form requirements, can be easily post-tensioned. The process of post-tensioning incorporates the installation of either bonded or unbonded tendons in preformed voids or ducts throughout the length of the member, or through a section of the member. After curing the member, strands are stressed and anchored against the hardened concrete.

Bonded tendons are installed in preformed voids or ducts and are made monolithic with the member and protected from corrosion by grouting after the stressing operation is completed. Unbonded tendons are protected against corrosion by a properly applied coating of galvanizing, epoxy, grease, wax, plastic, bituminous or other approved material, and are carefully cast in concrete in a sheathing of heavy paper or plastic. Unbonded tendons are connected to the member only through the anchorage hardware, which should be sized and designed in accordance with ACI 318.

The strand (tendon) most frequently used in architectural precast post-tensioned concrete is called the monostrand. Although monostrands can be fabricated to be grouted, they are usually coated with grease and covered with paper or plastic. Thus, they are typically used in the unbonded condition. If friction is exceptionally high due to length or curvature of the tendon, a strand coated with teflon and encased in a plastic tube is available. These monostrands have a low coefficient of angular friction ($\mu = 0.03$ to 0.05). Anchorages and pocket formers should be rigidly attached to the forms to prevent intru-

Standard

5.4.2 Details and Positions for Ducts

Ducts for post-tensioning tendons shall be constructed of flexible or semi-rigid metal or corrugated HDPE/polypropylene tubing installed within the member. Tendons which are not to be bonded by grouting may be installed in ducts of plastic, or other material. Metal ducts shall be of a ferrous metal and may be galvanized. Aluminum or PVC shall not be used for ducts.

The alignment and position of ducts within the member shall be controlled. The trajectory of ducts shall not depart from the curved or straight lines shown on the design drawings by more than 1/2 in. (13 mm) per 10 ft (3 m). For curved members, the tendons, and consequently the ducts, shall be placed on or symmetrically about the axis of the member that is parallel to the direction of the curvature. The position of ducts with respect to the thickness of the member, especially at critical locations shall be maintained within a dimensional tolerance consistent with the size and usage of the members, with a maximum variation from specified position of ±1/4 in. (±6 mm) or 1/8 in. per foot of depth, whichever is smaller.

The alignment of ducts shall be such that tendons are free to move within them and, if grouting is to be used, area shall be sufficient to permit free passage of grout. The inside diameter shall be at least 1/4 in. (6 mm) larger than the nominal diameter of single wire, bar or strand tendons; for multiple wire or strand tendons, the inside cross-sectional area of the duct shall be at least twice the net area of the prestressing steel.

Ducts installed in members prior to casting the concrete shall be of such construction that they will not admit concrete or mortar during casting. Ducts or duct forms shall be so supported and fastened that they will maintain their positions during casting and compaction of concrete. Joints between duct sections shall be adequately coupled and taped to maintain geometry and prevent concrete paste intrusion during casting. After placing of ducts and reinforcement and forming is complete, an inspection shall be made to locate possible duct damage. All holes, openings, or excessive dents shall be re-

Commentary

sion of cement paste into the anchorage cavity. Ties between the sheathed tendon and support steel should not be so tight as to cause visible deformations (indentations) in the sheating.

C5.4.2 Details and Positions for Ducts

Materials commonly used for formed ducts are 22 to 28 gauge galvanized or bright spirally wound or longitudinally seamed steel strip with flexible or semi-rigid seams.

Although most ducts are formed using metal tubing, occasionally collapsible or inflated rubber tubes that can be removed after the concrete has hardened are used to form a void in the member. This would not be a preferred method if grouting were to be utilized, due to the lack of composite action between the cylinder left by the void and the grout placed in the void. For grouted tendons a corrugated HDPE or polypropylene duct may be used if the materials meet the appropriate Post-Tensioning Institute recommendations.

Short kinks or wobbles in alignment will result in appreciable increases in friction during tensioning.

Tendons may be installed in the ducts either prior to or subsequent to placing concrete. In general, it is preferable to place the tendons subsequent to the concreting operation so that water and grout can be blown or cleaned from the duct to avoid blockage of the duct.

Standard

paired prior to concrete placing.

All ducts shall have grout openings at both ends. Grout openings and vents shall be securely anchored to the duct and to either the forms or to reinforcing steel to prevent displacement during concrete placing operations.

5.4.3 Friction in Ducts

The jacking force necessary to provide the required stress and overcome frictional force shall be indicated. Production documents shall also show the techniques to be observed in jacking, which may consist of over-jacking and over-elongation followed by a reduction of load for seating the anchorages or of jacking from both ends.

Maximum jacking force shall not exceed the applicable limits in ACI 318.

5.4.4 Tensioning

A schedule indicating the minimum jacking strength and a sequence of tensioning tendons to keep stresses within predetermined limits of symmetry about the axis of the member shall be established and shown on the production drawings. The concrete compressive strength shall be determined from test cylinders.

A minimum initial force of 10 percent of the jacking force shall be applied to the tendon to take up slack and to provide a starting point for elongation measurement. The jacking force shall then be applied, including any overload and release that may be called for in the procedure. The rate of application of the force shall be in accordance with the post-tensioning manufacturers' recommended procedure.

Final force applied to tendon and actual elongation measured shall check the theoretical values within 7 percent and shall agree with each other within 7 percent. If stressing is not achieved within this tolerance, then procedures shall be altered until tolerance limits are observed. For post-tensioned tendons the stress at the end anchorages, immediately after tendon anchorage lock-off shall not exceed $0.70 f_{pu}$.

5.4.5 Anchorages

Anchorage devices for all post-tensioning systems shall be aligned with the direction of the axis of the

Commentary

C5.4.3 Friction in Ducts

Friction on the post-tensioning tendon is due to length and curvature of the ducts. The length effect is the amount of friction that would be expected in a straight tendon due to minor misalignment (wobble of the duct). The curvature effect results from friction due to the prescribed curvature of the duct. Both components of this friction are proportional to the respective coefficients of friction between the tendon and the side of the duct. Coefficients and constants to be used for computing frictional effect have been established by research for all duct and tendon combinations in common usage.

C5.4.4 Tensioning

Post-tensioning in plant produced members is generally in short lengths so elongation is usually a small value which places added emphasis on carefully obtaining accurate readings.

For post-tensioned tendons shorter than 25 ft (7.6 m), special stressing methods and elongation measurement methods are required.

An accurate gauge is a necessity for unbonded tendons since stressing the strands by calculating the elongation would be difficult because of the elaborate strand configurations (multitude of tendon curvatures) and losses which occur during the jacking process.

The actual elongation of the unbonded strand should be checked against the theoretical elongation to ensure that the strand is entirely stressed. The strand may become bonded or kinked or the anchor may not be working properly preventing the strand from being fully stressed.

Sample calculations for tensioning setups are shown in Appendix H.

C5.4.5 Anchorages

For unbonded systems, the anchorage provides the only transfer point from tendon to member; therefore, it is critical that the anchorages be capable of 95 percent of the ultimate devel-

Standard

tendon at the point of attachment; concrete surfaces against which the anchorage devices bear shall be perpendicular to the tendon axis. Anchorage losses, due to seating loss or other causes, shall be measured accurately and compared with the assumed losses shown in the post-tensioning schedule and shall be adjusted or corrected in the operation when necessary.

The connections attaching the anchorages to the form shall be sufficiently rigid to avoid accidental loosening during concrete placement. The anchorage area shall be sealed immediately after the tendons or strand are post-tensioned. Minimum concrete cover for the anchorage shall not be less than the minimum cover to the reinforcement at other locations.

Plastic pocket formers used as a void form at stressing anchorages shall prevent intrusion of concrete or cement paste into the wedge cavity during concrete placement. Pocket formers shall be coated with grease prior to insertion to help prevent concrete leakage into the anchorage and to aid in their removal during form stripping.

5.4.6 Grouting

Ducts shall be blown free of water after curing of the concrete and provision made to keep water out of the ducts prior to grouting. To provide maximum protection to the tendons, grouting shall be performed within ten days after completion of the tensioning operation unless otherwise specified. If a delay is expected in grouting, a rust inhibitor can be applied to the tendon before placement in the duct.

Grout shall always be applied by pumping toward open vents. Grout shall be applied continuously under moderate pressure at one point in the duct until all entrapped air is forced out the open vent or vents. Vents shall not be closed until they discharge a steady stream of grout. Once all vents are closed, pumping shall continue until a steady pressure of 100 psi is maintained for ten seconds.

Thixotropic grouts shall be mixed with a shear-type mixer rather than a paddle mixer.

5.4.7 Sealing of Anchorages

Tendon anchorages shall receive a concrete or grout seal to provide the minimum cover required for the tendon material elsewhere in the structure. This seal shall be adequately covered for curing since shrinkage or contraction cracks will permit moisture pene-

Commentary

opment of the tendon.

Alignment of anchorages is critical for seating of tendons. Misalignment during casting can reduce effectiveness of anchorages.

C5.4.6 Grouting

Post-tensioning members which are to carry heavy fluctuating or dynamic loads or which are subject to frequent wetting or drying or severe climatic exposure should have the ducts containing the tendons pressure grouted following the completion of tensioning. Grouting is an important operation, serving to protect the tendons, relieve the anchorage of stress fluctuation, and increase the efficiency of the tendon in resisting ultimate moments.

C5.4.7 Sealing of Anchorages

Care should be exercised to protect end anchorages of tendons. Even with grouted tendons, the end anchorage is an integral part of the post-tensioned system.

Lack of proper protection allows an access point for moisture and a developing point for corrosion. If salts are present in the

Standard

tration. Low-shrinkage, non-metallic grout shall be chosen for anchorage pocket sealing.

If a concrete or grout seal cannot be provided, then the anchorage and tendon end shall be completely coated with a corrosion-resistant paint or other effective sealer. The anchorage and tendon end shall then receive a cover which will provide fire resistance at least equal to that required for the structure.

Commentary

moisture or water that accesses an anchorage, then corrosion is enhanced and often accelerated, making adequate end protection mandatory.

DIVISION 6 — QUALITY CONTROL

Standard

6.1 Inspection

6.1.1 Necessity for Inspection

To ensure that proper methods for all phases of production are being followed and the finished product complies with specified requirements, inspection personnel and a regular program of inspecting all aspects of production shall be provided in all plants. Inspectors shall be responsible for the monitoring of quality only and shall not be responsible for or primarily concerned with production.

Every effort toward cooperation shall be observed between production personnel, who are responsible for quantity and quality, and inspection personnel, who are responsible for observation and monitoring quality.

6.1.2 Scope of Inspection

To establish evidence of proper manufacture and quality of precast concrete, a system of records shall be utilized in each plant which will provide full information regarding the testing of materials, tensioning, concrete proportioning, placing and curing, and finishing.

In general, the scope of quality control inspections to be performed in precast concrete plants shall include, but not necessarily be limited to the following:

1. All required plant testing of materials for acceptance prior to initial placement and daily check testing for quality maintenance.

2. Mix design for all concrete and required concrete testing.

3. Inspection of molds and new set-up changes prior to placement of concrete. The plant shall prepare its own list of items to be checked as part of the pre-pour inspection, and emphasis shall be on items that cannot readily be checked after concrete placement.

4. Checking of blockout positioning, sealing strips, rustication strips, cast-in items, position and amount of reinforcement, and any other critical tolerance items, as well as the proper securing of these items during placement of concrete.

Commentary

C6.1 Inspection

C6.1.1 Necessity for Inspection

Pre- and post-pour inspections are useful for managing quality. Recurring defects require decisive action by management.

Plant management should give the quality control department sufficient time and resources to do an adequate job. Inspection operations should be so managed that production is not delayed as long as specified procedures are being followed. Many items must be checked during the pre-pour inspection and each type of panel (different mark numbers) has a different set of requirements. A plant's training program should include a definitive outline of items to be inspected.

C6.1.2 Scope of Inspection

To document the pre-pour inspection, quality control records should be identified with the same job number, mark number, and other information used to identify the product after inspection.

The post-pour inspection is frequently the last and sometimes the only opportunity to confirm that products were made in conformance with the shop drawings. The most important aspect of the post-pour inspection is the timeliness of the inspection. Post-pour inspections should be made as soon as practical after products are stripped from their molds. If a defect is evident or a mistake has been made and that defect or mistake is detected during the post-pour inspection, similar defects or mistakes in products yet to be cast can be prevented.

The number of persons required to perform inspecting services will vary with the size and scope of operations within the plant. It is important that a sufficient number of inspection personnel are available to carry out all prescribed tasks to maintain the thoroughness of inspections and tests. Assignments and responsibilities for all inspecting functions should be clearly defined and planned for in production scheduling.

Information gained through quality control inspection should be reviewed on a daily basis with production personnel. This review should be useful in identifying areas that may need production procedures reinforced or modified, or equipment that needs to be repaired or replaced.

Standard

5. Checking of molds and appurtenances for maintenance of tightness, dimensions and general quality with continued use.

6. Daily detailed inspection of batching, mixing, conveying, placing, compacting, curing and finishing of concrete.

7. Daily inspection of stripping from mold.

8. General observation of plant, equipment, working conditions, weather and other items affecting production.

9. Preparation of concrete specimens for testing and performing of tests for slump, air content, compressive strength, and other concrete tests.

10. Inspection of finish to make sure that the product matches the standard established by the approved project mockup or sample in color, texture, and uniformity. Finish defects, cracking, and other problems shall be reported and a decision made as to acceptance, repairs, or manufacturing change. Units which are damaged are to be recorded and marked.

11. Check finished product against approved shop drawings and approved samples to ensure that proper finishes are on all required areas, product measurements are correct, cast-in items are correctly located, panel is properly identified and marked, and that all measurements are within allowable tolerances.

12. General observations of storage area for proper blocking, methods for prevention of chippage, warpage, cracking, contamination or blocking stains, and any other items that may adversely affect the quality of the product.

13. Final inspection of product during loading for proper blocking and to detect stains, chips, cracks, warpage or other defects.

14. Inspection of products following any repair.

6.2 Testing

6.2.1 General

Testing shall be an integral part of the total quality control program. Testing for quality control of the precast concrete unit shall follow plant standards as well as standards required by the specifications for a particular project.

For control of concrete, testing of specimens, and the design and control of concrete mixes, each pre-

Commentary

C6.2 Testing

C6.2.1 General

Testing is necessary for internal plant quality control as well as quality control of the precast concrete unit. Testing should be directed towards maintaining a uniform level of plant standards.

Testing operations should be incorporated into plant operations to avoid unnecessary delays in production, and provide adequate product and process review.

Standard

cast concrete plant shall be equipped with adequate testing equipment, staffed with personnel trained in its use, or if the plant has contracted for quality control to be performed by an outside independent laboratory, that laboratory shall be accredited by the Cement and Concrete Reference Laboratory of the National Institute of Standards and Technology (National Voluntary Laboratory Accreditation Program). The laboratory shall conform to the requirements of ASTM E329 and the plant or independent laboratory shall meet the concrete inspection and testing section requirements of ASTM C1077.*

Specified properties of all materials in Divisions 3 and 4 shall be verified by appropriate ASTM standards performed by either the material supplier or the precaster.

In order to establish evidence of proper manufacture and conformance with plant standards and project specifications, a system of records shall be kept to provide full information on material tests, mix designs, concrete tests, and any other necessary information.

6.2.2 Acceptance Testing of Materials

Suppliers of materials shall be required to furnish certified test reports for cement, aggregates, admixtures, curing materials, reinforcing and prestressing steel, and hardware materials, indicating that these materials comply with the applicable ASTM standards, project specifications and plant standards.

1. **Cement.** If mill certificates are not supplied with each shipment, testing of cement is required. The mill certificate shall contain the alkali content in percent expressed as Na_2O equivalent. Mill certificates or test reports of cement shall be kept on file in the plant for at least 5 years after use.

---

* Titles for all standards and other documents referred to in the manual are given in Appendix G.

Commentary

C6.2.2 Acceptance Testing of Materials

In some instances, materials may not conform to nationally recognized specifications but may have a long history of satisfactory performance. Such nonconforming materials are permitted when approved by the architect/engineer when acceptable evidence of satisfactory performance is provided.

1. **Cement.** Mill test reports should be reviewed for changes from previous reports. Lower concrete strength should be expected from: lower cube strength; lower C_3S; lower fineness; higher % retained on No. 325 sieve (45 µm); and higher loss on ignition. Increase in total alkali may reduce concrete strength gain after 7 days and impair the strength-producing efficiency of water-reducing admixtures. Variation in a gray cement's color may in part be traced to a variable Fe_2O_3 content (a 2% variation being significant).

An additional report is available from cement producers that allows the concrete producer to evaluate cement strength uniformity (ASTM C917). The data will show 7 and 28-day cube strengths with 5 day moving averages and standard deviations. It is suggested that precasters routinely obtain these reports for the previous 6 to 12 months to monitor consistency of cube strengths.

If the tricalcium silicate (C_3S) content varies by more than 4%, the ignition loss by more than 0.5%, or the fineness by more than 375 cm^2/g, Blaine (ASTM C204), then problems in maintaining a uniform high strength

Standard

Characteristics of special cements not conforming to ASTM C150 shall be investigated prior to use to be certain that they do not exhibit undesirable attributes of high slump loss, strength retrogression, plateau-strength or other aberrations under typical casting and curing conditions.

2. **Aggregates.** Fine and coarse aggregates shall be regarded as separate ingredients. Aggregates shall conform to ASTM C33 or C330 except for grading and soundness. The grading requirements (only) of ASTM C33 or C330 shall be waived or modified to meet the special requirements of gap-graded face mixes and provide the advantages in having the backup mix as compatible as possible with the face mix. Sieve analysis in accordance with ASTM C136 shall be conducted on samples taken from the initial shipment received at the plant. Specific gravity, absorption, and petrographic analysis tests performed within the past 5 years shall be obtained from the supplier prior to the time of first usage or when a new lift or horizon in a quarry is utilized or there appears to be a change in quality of the aggregate.

The specific gravity factor of lightweight aggregate shall be determined in accordance with

Commentary

may result. Sulfate (SO_3) variations should be limited to ±0.20%.

It is good practice to keep a 10 to 15 lb cement sample (composite from 2 or 3 subsamples) in an airtight and moisture proof container with minimum air space over sample, until project acceptance, to check color and strength development, if necessary.

If problems have occurred with cement color variations, a visual check in sunlight of a cement sample for color conformity for each project should be made before allowing the cement to be loaded into the silo. A visual check of cement color should be made and compared to previous samples by placing the sample between two pieces of plate glass and taping the edges to hold in the cement. This is helpful in verifying that cement is from a standard mill source. Unannounced changes in mill sources may result in variable concrete properties, such as: air content, strength, setting and color.

2. **Aggregates.** Sampling of stockpiles or conveyor belts for aggregates should be in accordance with ASTM D75. Once a sample has been taken, the sample should be mixed and then quartered in accordance with ASTM C702. Once the sample has been reduced to a test size, testing should follow ASTM C136. Sieve analysis tests are required to ensure uniformity of materials received and to check consistency of gradation with the aggregate supplier's reported sieve analysis, taking into account expected changes in gradation that may be caused by rough handling in shipment.

Specific gravity and absorption of normal weight coarse aggregate should be determined according to ASTM C127 and for fine aggregate according to ASTM C128.

The specific gravity and absorption of an aggregate are used in certain computations for mixture proportioning and control, such as the absolute volume occupied by the aggregate. It is not generally used as a measure of aggregate quality, though some porous aggregates that exhibit accelerated freeze-thaw deterioration do have low specific gravities.

Petrographic analysis should be made in accordance with ASTM C295 to ensure that selected aggregates are durable, inert and free from iron sulfide (pyrite) and other deleterious materials. Petrographic examination may eliminate need for alkali reactivity tests. The frequency of testing will vary depending on the nature of the source of the aggregate.

For some relatively smooth surfaced, lightweight coarse aggregates, regular specific gravity and absorption proce-

Standard

procedures described in ACI 211.2, Appendix A — Pycnometer Method. The oven-dry loose unit weight (ASTM C29) of the lightweight aggregate shall be determined. A maximum 10 percent change in unit weight of successive shipments from sample submitted for acceptance tests is allowed.

Evaluation of aggregates for potential alkali-silica or alkali-carbonate reactions (excessive expansion, cracking or popouts in concrete) shall be based on at least 15 years of exposure to moist conditions of structures made with the aggregate in question, if available, or petrographic examination (ASTM C295) to characterize aggregates and determine the presence of potentially reactive components. If an aggregate is found to be susceptible to alkali-silica reaction using ASTM C295, it shall be evaluated further using ASTM C1260 and CSA A23.2-14A. Aggregates which exhibit ASTM C1260 mean mortar bar expansion at 14 days greater than 0.10 percent shall be considered potentially reactive. Aggregates further evaluated by CSA A23.2-14A that exhibit mean concrete prism expansion at one year greater than 0.04 percent, shall be considered potentially reactive. Aggregate sources exhibiting expansions no more than 0.04 percent and demonstrating no prior evidence of reactivity in the field shall be considered nonreactive. Reliance shall not be placed upon results of only one kind of test in any evaluation.

If an aggregate is judged to be susceptible to alkali-carbonate reaction using ASTM C295, it should be evaluated further for alkali-carbonate reaction in accordance with ASTM C586 or ASTM C1105.

ASTM D4791 shall be used to determine the percentage of flat and elongated particles in crushed coarse aggregate. Flat and elongated aggregate particles (slivers) shall be limited to a maximum of 15 percent by weight of the total aggregate. If aggregate is to be exposed in panel returns, the percentage shall be limited to 10 percent. ASTM C1252 (Method A) shall be used to evaluate angularity of fine aggregate.

Commentary

dures by ASTM C127 can be used; however, a lid is needed on the basket to confine floating pieces.

Lightweight aggregates should be ordered with specification restrictions. Uniformity of specific gravity and dry loose unit weight are important concerns for lightweight concrete. However, some sources do not consistently provide this material within a reasonable set of limits; therefore, adjustments in mix proportions may be required. Suppliers should forward gradation analyses and specific weight tests of the material with initial shipments. The specific weight test or weight of specific volume of load should be performed on each shipment so adjustments in batching can be made as the material changes in specific gravity from that assumed in the mix design.

Whenever possible, aggregates should be evaluated from their service records, taking into account the alkali contents of the cements used, whether the aggregate was used alone or in combination with another aggregate, and the exposure conditions and age of the concrete.

One generally accepted definition of a flat particle is one in which the width or length exceed the thickness by some ratio, usually 3:1. An elongated particle is one where the length exceeds the width by some ratio, usually 3:1. Flat pieces and elongated pieces (slivers) will produce irregular and non-uniform finishes when exposed and do not hold well in the concrete matrix during high pressure washing. Rough-textured, angular, elongated particles require more water to produce workable concrete than do smooth, rounded, compact aggregates. Hence, aggregate particles

Standard

These requirements may be waived if performance testing demonstrates satisfactory results.

Tests for deleterious substances and organic impurities shall be done at the start of a new aggregate supply and annually thereafter, unless problems are encountered requiring more frequent testing. Deleterious substances in aggregates shall be limited to the allowances given in ASTM C33 for exposed architectural concrete located in severe weathering regions with the following exceptions: (1) fine aggregate shall not exhibit a color darker than Organic Plate 1 when tested for organic impurities in accordance with ASTM C40; and (2) clay lumps and friable particles in fine aggregate shall be limited to maximum of 1 percent and in coarse aggregate to a maximum of 0.25 percent.

Coarse aggregates may occasionally contain particles with an iron sulfide content that results in unsightly stains. Since this aggregate could meet the staining requirements of ASTM C330, the requirements shall be tightened. If ASTM C295 indicates the presence of iron sulfides, then aggregates tested by ASTM C641 shall show a stain index less than 20.

Unless all aggregate is stockpiled at the beginning of a project, a sample of the approved aggregate for exposed surfaces shall be maintained until all units are accepted by the architect. As shipments of aggregates are received, a visual inspection shall be made such that the general appearance of the material can be compared with the approved aggregate sample.

3. **Water.** Water shall be potable or chemically analyzed when a private well or non-potable water is used in concrete mix. Except for water from a municipal supply, an analysis of the water shall be on file at the plant, be updated annually and be clearly related to the water in use.

    Mortar cubes made in accordance with ASTM C109 using non-potable or questionable mixing water shall have 7-day strengths equal to at least 90 percent of the strengths of companion specimens made with potable or distilled water. Time of set (ASTM C191) for mortar made with questionable water may vary from one hour earlier to 1-1/2 hours later than control sample

Commentary

that are angular require more cement to maintain the same water-cement ratio. In addition, long, slivery aggregate pieces produce concrete difficult to adequately consolidate because of aggregate interlock.

Many times friable particles or clay balls in aggregate, which are detected by ASTM C142, are weakened on wetting and may degrade on repeated wetting and drying.

Precast concrete units, although not normally exposed to salting or intense freezing and thawing, may be exposed to strong wet-dry cycling. Wet-dry sensitive coarse aggregates may crumble and such crumbling may be noticeable even if aggregate is used in small quantities.

Staining due to iron sulfides generally becomes noticeable at a later date due to moisture and oxidation from exposure to the atmosphere.

Standard

(made with municipal or distilled water). Water resulting in greater variations shall not be used.

4. **Reinforcing Steel and Prestressing Materials.** Plant testing of reinforcing steel, welded wire reinforcement, or prestressing materials shall not be required if mill certificates and coating reports are supplied. Mill certificates of reinforcing steel, welded wire reinforcement, and prestressing materials in stock or in use shall be required and indicate that they meet the requirements of applicable ASTM specifications and ACI 318. Mill certificates shall be kept on file in the plant for at least 5 years after use. Certificates shall be obtained for each size and shipment and for each grade of steel. Certificates shall be obtained for each ten reels or coils of prestressing strand or wire in each size, and for each heat or at least for each shipment if less than ten reels or coils. Certificates for prestressing materials shall be kept on file for each size, 10-ton shipment, each heat, or each shipment if less than 10 tons. Incoming steel, wire and strand shall be examined for damage, excessive scaling or pitting.

When it is required to restrict the range in the chemical composition of steel to provide satisfactory weldability, conformance with these supplemental requirements shall be certified in writing by the supplier.

The in-plant review and monitoring of welded wire reinforcement shall include a periodic inspection as it is received to confirm that the styles conform to the required size and spacing specified. Spacing of the wires shall be within 1/4 in. (6 mm) of the desired spacing, and the resistance welds at intersections of wires shall have not more than 1 percent bro-

Commentary

4. **Reinforcing Steel and Prestressing Materials.** It should be a standard practice to review mill certificates from suppliers of reinforcing bars for conformance to the purchasing requirements and shop drawings. As reinforcing shipments are reviewed, bar size and grade should be confirmed. Based on the chemical analysis, the welding criteria (if welding is to be utilized for reinforcing steel) should be established and the reinforcing steel marked accordingly. If the reinforcing steel is weldable according to AWS D1.1 and D1.4, the preheat required will vary depending on the chemical analysis of the steel.

Upon receipt of a shipment, bars should be reviewed for their general condition. Bars are acceptable with a certain degree of mill scale as well as some rusting. Bars should not have excessive pitting or loss of section caused by rusting. If the rust is easily removed by either finger pressure or a pencil eraser and no significant pitting is observed, the bars are in conformance with acceptable standards.

Upon receipt of prestressing steel, the mill certificates from the supplier should be reviewed for conformance with the purchase requirements. The prestressing steel should be physically inspected for confirmation of the material size and that the reels or the shipments are properly tagged and identified with a mill certificate and identifying that it is low relaxation or stress relieved strand. These materials should be inspected for excessive rust, nicks, and kinks which can cause problems in tensioning of the material. Rusting is generally acceptable if the rusting is light and if pitting has not begun. If rust can be removed by finger pressure or by the use of a non-metallic pad, it generally is not a cause for rejection. It should be noted that there are certain states, California being one, which for publicly-funded projects require that strands be bright and shiny and contain essentially no rust when used. Suppliers and plant personnel need to be aware of these special requirements when ordering, shipping, and storing strand. Kinks or nicks in strands, bars or wire provide an area where stresses are concentrated, and breakage can occur. Material which is received with kinks or nicks should not be accepted.

Standard

ken welds. Additionally, if specific finish requirements are specified, such as galvanizing or epoxy coating, this shall be confirmed at the point of delivery.

The stress-strain curve of the prestressing steel shall be on record. Stress-strain curves shall be for material tested from heats used to produce reel packs and shall be referenced to those reel packs. Average, typical or generic curves are not acceptable.

In lieu of mill certificates, reinforcing steel shall be tested for its physical and chemical properties in accordance with ASTM A370 to verify conformance with the applicable ASTM specifications.

5. **Admixtures.** The manufacturer of the admixture shall certify that individual lots meet the appropriate ASTM requirements. All relevant admixture information with respect to performance, dosages, and application methods and limitations shall be on file at the plant. Air-entraining admixtures shall conform to the requirements of ASTM C260. Other admixtures shall conform to the requirements of ASTM C494, Types A, B, D, F and G or ASTM C1017. The supplier shall certify these admixtures do not contain calcium chloride. Fly ash or other pozzolans used as admixtures shall conform to ASTM C618. Metakaolin shall conform to ASTM C618, Class N requirements and silica fume to ASTM C1240.

Laboratory test reports submitted by the supplier of chemical admixtures, shall include information on the chloride ion content and alkali content expressed as Na_2O equivalent. Test reports are not required for air-entraining admixtures used at dosages less than 130 mL per 100 kg (2 fl oz per 100 lb) of cement or nonchloride chemical admixtures used at maximum dosages less than 325 mL per 100 kg (5 fl oz per 100 lb). Both the chloride ion and total alkali content of the admixture are to be expressed in percent by mass of cement for a stated or typical dosage of the admixture, generally in milliliters per 100 kg or fluid ounces per 100 lb of cement.

6. **Pigments and Pigmented Admixtures.** The supplier shall certify that pigments or other coloring agents comply with the requirements of ASTM C979.

Commentary

5. **Admixtures.** The proprietary name and the net quantity in pounds (kilograms) or gallons (liters) should be plainly indicated on the package or containers in which the admixture is delivered. The admixture should meet ASTM requirements on allowable variability within each lot and between lots and between shipments.

It is desirable to determine that an admixture is the same as that previously tested or that successive lots or shipments are the same. Tests that can be used to identify admixtures include solids content, specific gravity using hydrometer, infrared spectrophotometry for organic materials, chloride content using silver nitrate solution, pH, and others. Admixture manufacturers can recommend which tests are most suitable for their admixtures and the results that should be expected. Guidelines for determining uniformity of chemical admixtures are given in ASTM C494, C233 and C1017.

Normal setting admixtures that contribute less than 0.1 percent chloride by weight (mass) of cement are most common and their use should be evaluated based on an application basis. If chloride ions in the admixture are less than 0.01 percent by weight (mass) of cementitious material, such contribution represents an insignificant amount and may be considered innocuous.

6. **Pigments and Pigmented Admixtures.** Synthetic mineral oxide pigments may react chemically with other products used on the surface, such as surface retarders or muriatic acid, and should be tested for these reactions prior to use.

Standard

7. **Hardware and Inserts.** Plant tests shall not be required for hardware but certification shall be obtained for all steel materials and each different grade of steel to verify compliance with specifications. Inserts need not be plant tested if used only as recommended by the suppliers and within their stated (certified) capacities and application qualifications. Records shall be on file establishing working capacity of each kind and size of insert used for handling and/or connection corresponding to the actual concrete strengths when inserts are used, unless the manufacturer's load table indicates adequate capacity at a strength lower than the maximum strength at time of use. No extrapolation of the suppliers test data is permitted. In lieu of certification for hardware, six specimens of each size and material heat number of a steel item shall be tested in accordance with ASTM A370 to verify conformance with the applicable ASTM specification. For other hardware items information shall be on hand describing the material, its qualities and applications, including limitations.

8. **Stud Welding.** Headed stud and deformed bar anchor materials and base metal materials shall be compatible with the stud welding process. Suppliers of both materials shall provide physical and chemical certification on the products supplied. The certification test shall have been made within the six month period before delivery of the studs.

    When assemblies are produced outside the precast concrete plant, the vendor producing the assemblies shall test assemblies in accordance with the testing procedure in Article 6.2.3(9). The procedure shall be provided to vendor and written into the sales agreement requiring conformance. Vendor personnel shall maintain records on an hourly basis and these records shall be provided to the precast plant. In addition, random sampling shall be done for each production lot of assemblies received at the precast plant. One unit for each 50 assemblies received shall be selected and the stud weld(s) visually inspected, and one stud bend tested, in accordance with the procedures detailed in Article 6.2.3(9). Any failure of the visual inspection or the bend test shall require like testing on a random 10 percent sample of the production lot. Any failure within this 10 percent sample shall require inspection and bend testing of 100 percent of the production lot, or

Commentary

7. **Hardware and Inserts.** If suppliers load tables indicate adequate capacity at a concrete strength lower than the design strength, the insert capacity is satisfactory.

8. **Stud Welding.** The heads of anchor studs are sometimes subject to cracks or "bursts" during manufacturing. This is essentially a crack starting at the edge of the head and progressing toward the center. As long as the cracks do not extend more than half the distance from the edge to the shank, as determined by a visual inspection, the crack is not cause for rejection. These interruptions do not adversely affect the structural strength, corrosion resistance or other requirements of the headed studs.

    An assembly failure in service can be inelastic and occur without warning. Two failures in 10 percent of a production lot is cause for serious concern. In addition to testing the entire lot, the manufacturer should be consulted and their procedures and materials checked for conformance with standards.

    A "production lot" is any collection of like assemblies received in a single shipment. Separate shipments of the same assembly type constitutes multiple production lots.

Standard

replacement of the entire lot.

Substitution of reinforcing bars for deformed bar anchors shall not be allowed unless approved by the precast engineer.

9. **Curing Compounds, Form Release Agents, Surface Retarders and Weatherproofing Sealers.** Instructions for proper use and application shall be obtained from suppliers and kept on file at the plant for all such materials used in the plant. If membrane curing compounds are used to retain moisture in concrete, such materials shall conform to ASTM C309; if sheet materials are used they shall conform to ASTM C171. Any coating used should be guaranteed by the supplier or applicator not to stain, soil, or discolor the precast concrete finish or cause joint sealants to stain concrete. Consult manufacturers of both sealants and coatings or pretest before applying the coating.

10. **Concrete Mixtures.** Concrete mix proportions shall be established under carefully controlled laboratory conditions. For lightweight concrete mixes, representative cylinders shall be cast and cured under plant production conditions to demonstrate the strength and weight of the concrete produced. All concrete mixes shall be developed using the brand and type of cement, the type and gradation of aggregates, and the type of admixtures proposed for use in production mixes. If at any time these variables are changed, the mix shall be re-evaluated. This re-evaluation may include one or more of the following concrete properties: (1) color, surface texture, or aggregate exposure; (2) air content or durability; or (3) strength (selected tests at appropriate ages).

    Records of all concrete mixes used in a plant and their respective test results shall be on file. Acceptance tests for concrete mixes shall include:

    a. **Compressive Strength.** Standard test specimens [6 x 12-in. (150 x 300 mm) or 4 x 8-in. (100 x 200 mm) cylinders or 4-in. (100 mm) cubes] shall be made and cured in accordance with ASTM C192 and tested in accordance with ASTM C39. Test specimens using 4-in. (100 mm) cylinders or cubes are permitted providing proper and proven correlation with the standard 6 x 12-in. (150 x 300 mm) test cylinder is available. In addition to the 28-day tests, compression tests

Commentary

Weldability characteristics of reinforcing bars are usually different, and strengths may not be the same.

10. **Concrete Mixtures.** Casting of a mix in a critical part of an actual mold is often advisable for checking a mix under production conditions.

    a. **Compressive strength.** Typically, the 28-day design strength established for face mixes and backup concrete should be not less than 5000 psi (34.5 MPa).

Standard

shall be made at the time of stripping the production unit from the mold to determine if stripping strength requirements have been met.

b. **Absorption.** The maximum water absorption for normal weight concrete (150 lbs. per cu. ft.) (2403 kg/m³) facing mixes at 28 days, shall not exceed 6 percent by weight. Alternatively, absorption expressed by volume shall not exceed 14 percent. The absorption of the facing mix of a continuous production run shall be verified every 6 months, for each new project, and whenever the materials and/or production methods are modified.

c. **Slump.** Slump tests shall be made in accordance with ASTM C143.

d. **Unit Weight.** Unit weight shall be tested in accordance with ASTM C138 or C567.

e. **Air Content.** Air content in concrete shall be measured in accordance with ASTM C173 or C231 as applicable.

Commentary

b. **Absorption.** A water absorption test of the proposed facing mixes may provide an early indication of weathering properties of the concrete (rather than durability).

Samples for testing. Three 4 x 8-in. (100 x 200 mm) cylinders or 4-in. (100 mm) cubes should be cast with concrete from three different batches from each of the mixes being tested. If possible, samples should be cast in containers made from the mold material intended for the actual production unit. Test samples should be consolidated, cured and finished similar to the products they represent. Test samples should be clean and free from any parting or form release agent or any sealer.

Test procedure. Specimens should be tested after 28 days in accordance with ASTM C 642 except procedures described in Sections 5.3 and 5.4 of the ASTM test are not required. The percentage absorption is the average absorption of the three specimens. This figure may be transformed to volume percentage based on the specific weight of the concrete tested.

A comparison of water quantities absorbed after a given time, for instance 15 minutes, may be determined from Rilem Test Method II.4, Water Absorption under Low Pressure (Pipe Method). This test procedure can be used to assess the effectiveness of a sealer. However, a correlation to ASTM C 642 has not been developed, therefore acceptance criteria have not been established for the Rilem Test.

c. **Slump.** The standard slump test may be inadequate as a measure of the workability of concretes with high proportions of coarse aggregate.

e. **Air Content.** The volumetric method of checking air entrainment (ASTM C173) may be used on any type of aggregate, whether it be dense, cellular or lightweight. The pressure method (ASTM C231) gives excellent results when used with concrete made with relatively dense natural aggregates for which an aggregate correction factor can be determined satisfactorily. It is not recommended for use on concretes made with lightweight aggregates, air-cooled blast furnace slag, or aggregates of high porosity. It also may not work properly on very harsh or low-slump mixtures.

Since architectural precast concrete units are generally erected in an above-grade vertical position, which is a moderate environment, air contents as low as 3 to 6 percent in the concrete or 7% air in the mortar faction appear to provide the required durability. Air-entrainment levels no higher than necessary are preferred,

Standard

6.2.3 Production Testing

Production testing shall be directed towards maintaining production and product uniformity by routine testing of materials and concrete to ensure that they are consistent with supplier's reported data or established requirements.

1. **Aggregates.** A sieve analysis (ASTM C136) and unit weight test (ASTM C29) shall be conducted in the plant with test samples taken at any point between and including stockpile and batching hopper. Such tests shall be carried out for each aggregate type and size at least once every 2 weeks or for each of the following aggregate volumes, where usage in a 2-week period exceeds such volume:

   a. Aggregates used in architectural face mixes
   .... 40 cu. yds. (31 m³)

   b. Fine aggregates used for backup mixes
   .... 200 cu. yds. (150 m³)

   c. Coarse aggregates used for backup mixes
   .... 400 cu. yds. (310 m³)

   Each shipment of aggregate shall be visually compared with the approved aggregate sample.

Commentary

since the compressive strength of concrete is reduced by approximately 5 percent for each 1 percent of entrained air (when the water-cement ratio is held constant). Strength reductions tend to be greater in mixes containing more than 550 lb. of cement per cubic yard (326 kg/m³). Since most precast concrete mixes contain a high cement factor, relatively high reductions in strength may be anticipated with high levels of air entrainment. Also, as compressive strengths increase and water-cement ratios decrease, air-void parameters improve and entrained air percentages can be set at the moderate exposure limits.

The addition of normal amounts of an air-entraining agent to harsh gap-graded facing mixes will improve the workability and increase resistance to freezing and thawing even though only a small amount of air is usually entrained.

C.6.2.3 Production Testing

Quality-control charts displaying production test results should be used to uncover unanticipated variations in materials, batching, mixing, curing, and testing concrete. The primary objective of quality-control charts is to test whether or not a process is in statistical control. Control charts are valuable in visually presenting the data in a manner where variation can be readily seen. These charts can provide information on whether a problem exists in a concreting operation; however, quality-control charts may not locate where the variability is occurring. Quality-control charts do provide clues on where to look for process variability. Quality-control charts provide the benefits of (1) limiting defective batches, (2) fewer rejected batches, and (3) better overall quality.

For further information on the use of control charts refer to *ACI Manual of Concrete Inspection, SP-2* and Gebler, Steven H., *Interpretation of Quality Control Charts for Concrete Production"* ACI Materials Journal, July/August, 1990, pp. 319-326.

1. **Aggregates.** Sampling preferably should be from conveyors or from the discharge opening of bins. Stockpiles are most difficult to sample properly and should be avoided as sample sources, if possible. The most representative sample possible is that from a conveyor belt. For fine aggregate take scoopfuls from the belt until a bucketful is obtained, from which the test sample can be split or quartered. For coarse aggregate, take samples from the belt only if it is practicable to stop it while all material on a short length of belt is removed. If arrangements cannot be made to stop the belt, or if there are no belt conveyors at the plant, other means must be used. The next best method is to take the entire momentary discharge of coarse aggregate from a chute or bin gate. Take at least several cubic feet of material and quarter the test sample from this amount. Such samples are most representative when it is possible to quarter them from material taken from the first, middle, and last of the material to be tested.

Close control over gradation of aggregates is essential to minimize variations in surface texture and color in the fin-

Standard

Moisture tests are not required for bagged aggregates stored indoors. Surface moisture in bulk aggregates shall be evaluated and compensated for in all concrete proportioning. Moisture content shall be determined by drying (ASTM C566), by a meter that measures moisture by the pressure of chemically generated gas, by an electric probe with moisture indicated by the resistance between electrodes, by microwave energy absorption, or other devices calibrated against ASTM C566..

Either the moisture meter or electric probe are satisfactory for continuous moisture determination provided they are calibrated against the drying method (ASTM C566). If moisture meters are not used, the free moisture shall be determined at least daily, or at any time a change in moisture content becomes obvious and appropriate corrections shall be made.

2. **Concrete Strength.** During production, concrete shall be sampled and specimens made in accordance with the following specifications except as modified herein:

    ASTM C172 — Sampling Fresh Concrete.

    ASTM C31 — Making and Curing Concrete Test Specimens in the Field.

    Each sample shall be obtained from a different batch of concrete on a random basis, avoiding any selection of the test batch other than by a

Commentary

ished product. Aggregates should be handled and stored in a way that minimizes segregation and degradation and prevents contamination by deleterious substances. Dry-rodded unit weight of aggregate is important for mix design while variations in dry-rodded unit weight may indicate a change in gradation, specific gravity or particle shape.

It is good practice to maintain a running average on from 5 to 10 previous gradation tests, dropping the results of the oldest and adding the most recent to the total on which this average is calculated. These averages can then be used to make necessary adjustments to mix proportions.

For coarse aggregate, determine the loose density and then calculate void content based on the specific gravity of the aggregate. If the number changes from the initial supply have the aggregate supplier investigate the variation.

Compensation for surface moisture is particularly important for face mixes where the amount of fine aggregates batched by weight may vary enough to seriously affect the color and texture of the finished face. The free moisture on aggregates affects net aggregate weights as well as the amount of water added to the batch and may cause either over- or under-yielding of concrete mixes. It is recommended that weighing hoppers be equipped with properly maintained moisture meters periodically calibrated to detect changes of 1 percent in the free moisture content of fine and coarse aggregates so corrections can be made and mixes adjusted at any time. Readings from moisture metering devices based on conductivity will vary with the density of the aggregates and are not recommended for lightweight aggregates. Determination of moisture content by drying is time consuming and not necessarily accurate for practical concrete proportioning as it tests only an isolated sample.

2. **Concrete Strength.** Testing of concrete strengths by using test specimens is a critical part of the quality control program. In addition, properly correlated rebound hammer or other nondestructive tests may be used to indicate stripping strength.

Samples for strength tests must be taken on a strictly random basis if they are to measure properly the acceptability of the concrete. A predetermined sampling plan (chance approach) should be set up before the start of production

Standard

number selected at random before commencement of concrete placement.

The size of specimens made and cured in accordance with ASTM C31 are modified to permit use of either 6 x 12-in. (150 x 300 mm) or 4 x 8-in. (100 x 200 mm) test cylinders or 4-in. (100 mm) cubes. Test specimens using 4 x 8-in. (100 x 200 mm) cylinders or 4-in. (100 mm) cubes shall be considered acceptable, provided that proper correlation with standard 6 x 12-in. (150 x 300 mm) test cylinders is available.

Maximum size of aggregate in 4 x 8-in. (100 x 200 mm) cylinder or 4-in. (100 mm) cubes should not exceed 1 in. (25 mm). If larger sized aggregate is contained in the concrete, the compressive strength shall be measured using standard 6 x 12-in. (150 x 300 mm) cylinders. Special cube size may be used when they more adequately represent particular products, if correlation to standard 6 x 12-in. (150 x 300 mm) cylinders is provided.

Four compression specimens shall be made daily for each individual concrete mix (whether facing or backup mix), or for each 40 cu. yds. (31 m^3) of any one mix where the daily consumption exceeds this volume.

The required average strength of the concrete

Commentary

by establishing the intervals at which samples will be taken. The intervals may be set in terms of either time elapsed or yardage placed. The choice of times of sampling or the batches of concrete to be sampled must be made on the basis of chance alone within the period of placement in order to be representative. If batches of concrete to be sampled are selected on the basis of appearance, convenience, or other possibly biased criteria, statistical concepts lose their validity. Obviously, not more than one test should be taken from a single batch, and water should not be added after the sample is taken.

Development of a correlation curve allows the use of a new or unapproved testing method after correlating that method to an approved method of testing. Correlation testing should be done annually as a minimum and at the start of a new mix design and should include the following:

1. A correlation curve should be established for each combination of concrete mix design, curing procedure, and age of test.
2. A minimum of 30 tests should be used for each correlation curve.
3. Test results should fall within the 95 percent confidence limit of the correlation curve.

If values cannot be obtained during testing which are consistently within the 95 percent confidence limit, then a valid correlation curve cannot be established. It is recommended that the specific procedure, specimen size or test method being considered not be used, since it cannot be related to the standard with the required degree of confidence.

Although 4-in. (100 mm) diameter cylinders or cubes tend to test slightly higher than 6-in. (150 mm) diameter cylinders, the difference is usually insignificant. In the absence of adequate test data for a correction factor, a factor of 80 percent of the strength should be used as a temporary conservative measurement to convert to standard (6 x 12 in.) (150 x 300 mm) cylinder strength for the same concrete.

Preparation and testing of cubes should be nearly as consistent with the appropriate requirements for cylinders as possible, with the exception that the concrete be placed in a single layer of 4 in. (100 mm). Rodding or external vibration methods would then proceed as outlined in the ASTM designations. Internal vibration should not be applied to the consolidation of cubes.

Standard

shall be selected in accordance with Chapter 5, ACI 318.

Cylinder molds shall be kept clean and free from deformations and conform to requirements of ASTM C470.

Test specimens shall be made as near as possible to the location where they will be cured and shall not be disturbed in any way from 1/2 hour after casting until they are either ready to be stripped or tested. Specimens shall be protected from rough handling at all ages.

Test specimens shall be cured with and by the same methods as the units they represent up to the time of stripping or detensioning from form or mold. Cylinders stored next to a product shall have their curing conditions verified as similar to the product. In lieu of actual curing with the member, cylinders may be cured in curing chambers correlated in temperature with the product they represent. In such a case, the correlation shall be constantly verified by use of recording thermometers in the curing chambers and comparison with the temperature records of the product, and by use of the same methods of moisture retention for curing chambers and casting beds.

After stripping of the unit, test specimens shall be removed from their molds and stored in a moist condition at 73.4 deg. F ± 3 deg. F (23 ± 1.7 deg. C) until the time of testing.

Unless specimen ends are cast or ground to within 0.002 in. (0.05 mm) of a plane surface, they shall be capped prior to testing or unbonded caps (elastomeric pads) may be used in accordance with ASTM C1231. Capping procedures shall be as specified in ASTM C617 except that with fast setting sulfur compounds, especially manufactured for capping, compression testing may be performed 1/2 hr. after the caps have been in place. The casting temperature of capping compounds shall be controlled. Thermostatically controlled heating pots shall be used.

Commentary

Molds for making test specimens should be in accordance with applicable requirements of ASTM C31 and C470.

The strength of a test specimen can be greatly affected by jostling, changes in temperature, and exposure to drying, particularly within the first 24 hours after casting. Thus, test specimens should be cast in locations where subsequent movement is unnecessary and where protection is possible. Cylinders should be protected from rough handling at all ages. Because of the danger of producing cracks and weakened planes, concrete with slumps less than about 1 in. (25 mm) should not be moved even in the first 15 to 20 min. Concrete in cylinders may be consolidated by rodding or by vibration as specified in ASTM C31. Any deviations from the requirements of ASTM C31 should be recorded in the test report. If vibrators are used, techniques should be developed to preclude segregation.

Standard

Testing of specimens to determine compressive strength shall be performed in accordance with ASTM C39. The strength of concrete at any given age shall be determined as the average of two specimens except one specimen can be used to determine stripping or stress transfer strength as production progresses. If either specimen shows definite evidence, other than low strength, of improper sampling, molding, handling, curing or testing, it shall be discarded, and the strength of the remaining cylinder shall be considered the test result.

The strength level shall be considered satisfactory if both of the following requirements are met:

1. The average of all sets of three consecutive strength tests equal or exceed the specified 28-day strength.
2. No individual strength test (average of two cylinders) is more than 500 psi (3.4 MPa) below the specified strength.

Non-destructive tests can be useful tools to supplement, but not replace, cylinder tests, except as noted. They can serve to give a comparative or qualitative evaluation of concrete strengths. They may serve to determine stripping or transfer strengths when cylinders have been damaged or have all been used and to indicate shipping strength.

Non-destructive test methods shall be acceptable provided the following conditions are met:

1. A correlation curve is established for each combination of concrete mix design, curing procedure, and age of test.
2. A minimum of 30 tests is used for each correlation curve.
3. Test results fall within the 95 percent confidence limits of the correlation curve.
4. Correlation curves are established for each test instrument, even of the same type.

Where concrete strengths are to be evaluated by an impact hammer in accordance with ASTM C805, at least 3 sets of readings (test areas) shall be taken along the member. The area to be tested shall have a uniform surface produced by stone rubbing. Care shall be taken not to take readings on an isolated piece of coarse aggregate or on any surface imperfection or directly over reinforcing steel or other steel near the surface.

Commentary

Non-destructive testing is testing generally performed on the product rather than a sample and does not destroy the product or area tested. The most common method is the impact hammer. For all such testing, the most important criterion is correlation testing back to cylinders to provide a correlation factor. Such testing should be done at the beginning of use of non-destructive testing and annually thereafter.

Standard

The compressive strength for concrete in the area tested shall be taken from the point on the calibration curve corresponding to the average reading; and the strength of the unit shall be taken as the average of the strengths indicated at the three test areas.

If the compressive strength of a unit is questionable because test results were more than 500 psi below specified strength (Article 5.5.1), at least three cores shall be taken from each unit considered potentially deficient. Test cores shall be obtained, prepared and tested in accordance with ASTM C42. If the panel represented by the cores will be dry under service conditions, cores shall be air dried (at room temperature with the relative humidity less than 60 percent) for 7 days, and shall be tested dry. Concrete in the unit represented by the core tests shall be considered structurally adequate if both of the following requirements are met:

1. The average strength of three cores is equal to at least 85 percent of the specified strength
2. No single core is less than 75 percent of the specified strength.

3. **Slump.** Slump tests for each concrete mix design shall be made in accordance with ASTM C143 at the start of operations each day, when making strength test specimens, whenever consistency of concrete appears to vary, and every second or third air content test.

The following tolerances shall be allowed for individual batches provided the slump variation does not affect appearance or other qualities of the concrete beyond that allowed in the specifications:

Slump, where specified as "maximum" or "not to exceed," for all values . . . . +0 in.

Specified slump 3 in. (75 mm) or less . . .
. . . .+0, -1-1/2in. (+0, -38 mm)

Specified slump more than 3 in. (75 mm)

Commentary

If it is necessary to drill cores, their location should be determined by the precast engineer to least impair the strength of the structure and exposed surface finish. Core holes can often be adequately patched without damage to the appearance or structural integrity of the element.

When possible, cores should be drilled so that the test load is applied in the same direction as the service load. Horizontally (in relation to casting) drilled cores may be up to 15 percent weaker than vertically drilled cores. Cores should be drilled with a diamond bit to avoid an irregular cross section and damage from drilling. If possible, cores should be drilled completely through the member to avoid having to break out the core. If the core must be broken, wooden wedges should be used to minimize the likelihood of damage and allow 2 extra in. (50 mm) of length at the broken end to permit sawing off ends to plane surfaces before capping. If any core shows evidence of damage prior to testing, it should be replaced.

The inclusion of reinforcing steel in the cores may either increase or reduce the strength. The variation tends to be larger for cores containing two bars rather than one. The cores, therefore, should be trimmed to eliminate the reinforcement provided a length to diameter ratio of 1.00 or greater can be attained.

3. **Slump.** The slump test is a measure of concrete consistency. For given proportions of cement and aggregate without admixtures, the higher the slump, the wetter the mixture. Slump is indicative of workability when assessing similar mixtures. However, it should not be used to compare mixtures of totally different proportions. When used with different batches of the same mixture, a change in slump indicates a change in consistency, aggregate grading or moisture content, cement or admixture properties, amount of entrained air or temperature. Thus, slump test values are indicative of hour-to-hour or day-to-day variations in the uniformity of concrete.

Flowable concrete achieved by the incorporation of high range water reducers (HRWR) (superplasticizers), are difficult to control within tight tolerances at specified slumps of 7 in. (175 mm) or greater. In addition, it is difficult to accurately measure high slumps. Consideration should be given to eliminating a maximum slump requirement when a HRWR is used to achieve flowable concrete.

Standard

.... +0, -2-1/2 in. (+0, -63 mm)

Slump, when specified as a single value

Specified slump 4 in. (100 mm) or less . . . .
.... ± 1 in. (±25 mm)

Specified slump more than 4 in. (100 mm) . .
.... ± 1-1/2 in. (±38 mm)

Where range is specified there is no tolerance.

For superplasticized concrete . . . .
.... ± 2-1/2 in. (±63 mm)

4. **Air Content.** If an air-entraining admixture is used, the air content of concrete shall be measured in accordance with ASTM C173 or C231 as applicable. Air content shall be tested periodically during the daily operation with a minimum of one daily check per mix design or when making strength test specimens.

   Variations from the established value of air content shall not exceed 1-1/2 percentage points to avoid adverse effects on compressive strength, workability, or durability.

   A check on the air content shall be made when the slump varies more than ± 1 in. (25 mm), temperature of the concrete varies more than ± 10 deg. F (5 deg. C), a change in aggregate grading occurs, or there is a loss in concrete yield.

5. **Unit Weight.** Unit weight tests of concrete in accordance with ASTM C138 shall be carried out at least once per week for each mix design used regularly except for lightweight concrete which shall be tested daily in accordance with ASTM C567 to confirm batching consistency. When the nominal fresh unit weight varies from the established value by more than ±2 lbs. per cu. ft (±32 kg/m³) for normal weight concrete or ±2 percent for structural lightweight concrete, batch adjustments shall be made.

Commentary

4. **Air Content.** The volumetric method of measuring air content (ASTM C173) may be used on any type of aggregate, whether it be dense, cellular or lightweight. The pressure method (ASTM C231) gives excellent results when used with concrete made with relatively dense natural aggregates for which an aggregate correction factor can be determined satisfactorily. It may not work properly on very harsh or low-slump mixtures. With such mixtures, the application of pressure to the surface of the concrete may not result in the expected compression of the air in the void system. The volumetric method, ASTM C173, is not subject to this limitation and should produce accurate results on even the driest concrete.

   Although these tests measure only air volume and not air-void characteristics which can only be determined microscopically (ASTM C457), it has been shown by laboratory tests that these methods are generally indicative of the adequacy of the air-void system.

5. **Unit Weight.** The unit weight is a quick and useful measurement for controlling quality. A change in unit weight generally indicates a change in either air content or aggregate weight. When unit weight measurements (ASTM C138 or C567) indicate a variation of the calculated fresh unit weight from the laboratory mix design of more than 2 lbs. per cu. ft. (32 kg/m³) for normal weight concrete or ± 2 percent for structural lightweight concrete, the air content should be checked first to establish if the correct amount of air has been entrained. If air contents are correct, then a check should be made on the aggregates to make certain that the unit weight, gradation, moisture content or proportions used have not changed. Results of these checks generally will reveal the cause of the variations in

Standard

6. **Temperature of Concrete.** The temperature of freshly mixed concrete shall be measured in accordance with ASTM C1064 and recorded whenever slump or air content tests and compressive test specimens are made, at frequent intervals in hot or cold weather, and at start of operations each day.

7. **Air Temperature.** Ambient air temperature shall be recorded at the time of sampling for each strength test.

8. **Welding.** Quality control shall verify the welder's qualification, making certain that the proper electrodes (oven dry) are available and used, and that the preheat temperature indicating devices are at hand. Welder qualification shall be for the welding process, expected weld types and position of welds to be performed. As a minimum, welder shall be qualified for complete joint penetration groove welds and flare-groove welds.

The welder's qualification shall be considered as remaining in effect indefinitely unless: (1) the welder is not engaged in a given process of welding for which the welder is qualified for a period exceeding six months, or (2) there is some specific reason to question a welder's ability.

Personnel responsible for acceptance or rejection of welding workmanship shall be qualified. The following are acceptable qualification bases:

a. Current or previous certification as an AWS Certified Welding Inspector (CWI) in accordance with the provisions of AWS QC1.

b. Current or previous qualification by the

Commentary

unit weight of concrete and indicate what mix adjustments need to be made. After adjustments are made, the unit weight should again be measured.

The unit weight test results are used to calculate the volume or yield produced from known weights of materials and to calculate the cement content in pounds per cubic yard of concrete.

6. **Temperature of Concrete.** Temperature of fresh concrete affects a number of properties of concrete. Warm concrete sets faster than cool concrete. Warm concrete requires more water per cubic yard than cool concrete to produce the same slump. So for mixes of the same slump, unless more cement is used in the warmer concrete, the concrete will have a higher water-cement ratio. Warm concrete gains strength faster than cool concrete, but the strength at later ages may be lower than that of cool concrete. A knowledge of the temperature of fresh concrete permits the batch plant operator to adjust mixes. Higher concrete temperatures result in more air-entraining agent being needed to produce a given air content. Warm concrete tends to dry faster so curing of warm concrete is even more important than curing of cool concrete. Also important is maintaining a given minimum temperature during cold weather concrete operations, both during placement and during the time that protection is necessary.

8. **Welding.**

Standard

Canadian Welding Bureau (CWB) to the requirements of the Canadian Standard Association (CSA) Standard W178.2.

c. An engineer or technician who, by training, or experience, or both, in metals fabrication, inspection and testing, is competent to perform inspection of the work.

The qualification of the responsible personnel shall remain in effect indefinitely, provided they remain active in inspection of welded steel fabrication, unless there is specific reason to question the personnel's ability.

Inspectors shall have passed an eye examination with or without corrective lenses to prove: (1) near vision acuity of Snellen English, or equivalent, at 12 in. (305 mm); and (2) far vision acuity of 20/40, or better. Vision examination of all inspection personnel is required every three years or less if necessary to demonstrate adequacy.

Prior to welding, inspection shall include the following:

a. Reviewing welding drawings and welding procedure specifications.
b. Assuring that welding materials and consumables are in accordance with specifications.
c. Checking and identifying materials as they are received against specifications.
d. Checking storage of filler material.
e. Checking welding equipment to be used.
f. Checking weld joint preparations.
g. Checking for base metal discontinuities.
h. Establishing a plan for the recording of results.

Visual inspection during welding by weld operator shall include:

a. Quality of weld root bead.
b. Joint root preparation, such as slag removal, prior to welding the second side.
c. Preheat and interpass temperatures.
d. Sequence of weld passes.
e. Subsequent layers for apparent weld quality.
f. Cleaning between passes (use of a wire brush and chipping hammer to get rid of any slag).
g. Conformance with the applicable procedure, i.e., voltage, amperage, heat input, speed.

Commentary

Visual inspection guidelines are given in AWS B1.11 while radiographic and ultrasonic testing procedures and limits are given in AWS D1.1 and D1.4 and should be referenced, if needed. Radiographic testing is good and ultrasonic testing is poor in detecting volumetric discontinuities, such as porosity. Ultrasonic testing is good for detecting planar discontinuities, such as incomplete sidewall fusion, while radiographic testing can miss such discontinuities unless these are oriented parallel or near parallel to the radiation.

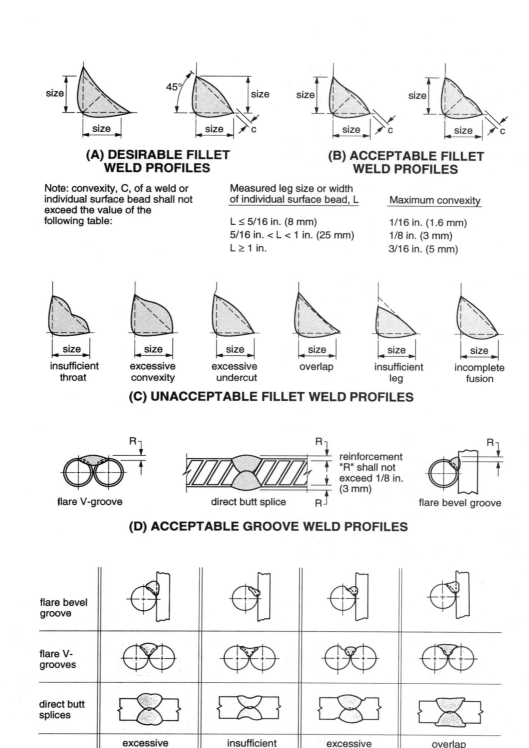

Fig. 6.2.3a. Welded Profiles.

Standard

The following items shall be checked on at least 10 percent of all assemblies after welding to determine the quality of the welds:

a. Geometric imperfections. The fillet weld faces shall be slightly convex or slightly concave as shown in Fig. 6.2.3a, A and B or flat, and with none of the unacceptable profiles exhibited in Fig. 6.2.3a, C.

b. Welds shall have no cracks in either the weld metal or heat-affected zone of the base metal.

c. There shall be thorough fusion between weld metal and base metal and between successive passes in the weld.

d. All craters shall be filled to the full cross section of the weld, except for the ends of intermittent fillet welds outside their effective length.

e. Welds shall be free from overlap.

f. For materials less than 1in. (25 mm) thick, undercut depth greater than 1/32 in. (1 mm) in the solid section of the bar or structural members shall not be allowed except at raised reinforcing bar deformation where 1/16 in. (1.5 mm) is allowed. For steel, refer to AWS D1.1 for the requirements of the specific structure type.

g. For reinforcing bars, the sum of diameters of piping porosity in flare-groove, and fillet welds shall not exceed 3/8 in. (10 mm) in any linear inch (25 mm) of weld and shall not exceed 9/16 in. (14 mm) in any 6 in. (150 mm) length of weld. For steel, refer to AWS D1.1 for the requirements of the specific structure type.

h. Incomplete joint penetration.

i. Slag inclusions.

j. Amount of distortion.

In addition to checking that required weld size, length, location and type are as indicated on the approved drawings, and that no welds are omitted, a check shall be made that no welds have been added without approval.

Weldments shall be checked after fabrication for brittleness by striking at least one out of every 50 pieces with a 3 lb (1.3 kg) hammer. Brittle weldments break under a hammer blow and if brittle weldment is found, all assemblies made using similar procedures become suspect and shall be checked for acceptance.

Commentary

b. Visual inspection for cracks in welds and base metal and other discontinuities should be aided by a strong light, magnifiers, or such other devices as may be found helpful.

c. Only incomplete fusion apparent during a visual inspection need be considered unless specifications require radiography or ultrasonic examination.

Size, length, and contour of welds should be measured with suitable weld-size gauges. Groove welds should be measured for proper reinforcement on both sides of the joint. Quality control should compare the welds with three-dimensional "workmanship samples" available from AWS. These are actual welded samples, or plastic replicas of welded samples that depict actual weld conditions.

Standard

Welds shall be corrected by rewelding or removal, in accordance with specified procedures.

9. **Stud Welding.** The stud-welding operator shall be responsible for the following tests and inspections to ensure that the proper set-up variables are being used for the weld position, stud diameter and stud style being welded. Testing is required for the first two studs in each day's production and any change in the setup such as changing of any one of the following: stud gun, stud welding equipment, stud diameter, gun lift and plunge, total welding lead length, or changes greater than 5 percent in current (amperage) and dwell time.

For studs welded in the down hand position, the operator shall at the start of each production period, weld two studs of each size and type to a production weld plate or to a piece of material similar to the weld plate in material composition and within ±25 percent of the production weld plate thickness. The test weld plate and production weld plate pieces shall be clean of any dirt, paint, galvanizing, heavy rust or other coatings which could prevent welding or adversely affect weld quality. These studs shall be visually inspected by the operator to see that a proper weld-fillet has formed. The weld-fillet (flash) may be irregular in height or width, but shall completely "wet" the stud circumference without any visual sign of weld undercut.

The test studs shall exhibit an after weld length measurement shorter than the before weld stud length. After weld length shall be consistent on both test welds and on all production welds. Typical length reductions for various stud diameters are shown in Table 6.2.3a.

After the test studs are allowed to cool, they shall be bent to an angle of approximately 30 deg. from their original axis by striking the studs with a hammer or placing a pipe or other suitable hollow device over the stud and manually or mechanically bending the stud. At temperatures below 50 deg. F (10 deg. C), the stud shall be bent slowly using the hollow device only. Threaded studs shall be torque tested in accordance with AWS D1.1 to a proof load of approximately 80 percent of their specified yield strength rather than bend tested.

Completion of the visual and mechanical tests listed above on both studs without evidence of failure in the weld zone constitutes acceptance

Commentary

9. **Stud Welding.** Testing of sample studs should be part of daily operations. If failures occur in plate stock to which a stud is welded, then requirements for plates should be reviewed by engineering.

At temperatures below 50 deg. F (10 deg. C), some materials for the stud and base materials lack adequate toughness to pass a hammer test.

Standard

Table 6.2.3a. Length reductions.

Stud Diameter, in. (mm)	Length Reduction, in. (mm)
$3/16$ (5) through $1/2$ (12)	$1/8$ (3)
$5/8$ (16) through $7/8$ (22)	$3/16$ (5)
1 (25) and over	$3/16$ to $1/4$ (5 to 6)

of the stud welding procedure and qualifies the process and the operator for production welding in the down hand position.

If either stud fails the visual, measurement or bend test inspection, the operator shall check the welding variables and make the necessary adjustments. Two additional studs shall be welded by the operator after adjustment and re-tested per the above procedure. Failure of either of the second set of test specimens shall cause the stud welding operation to be stopped and appropriate supervisory personnel notified.

Studs welded to positions other than down hand and studs welded to the heel of an angle or into the fillet of an angle shall be subject to the same pre-production tests and inspections except that ten (10) studs shall be welded and satisfactorily bend tested 90 deg. without failure prior to proceeding with production welding.

Weld procedure specifications (WPS) and procedure qualification records (PQR) shall be maintained by the appropriate supervisory personnel if required by the owner or engineer of record.

Pre-production test samples shall be identified and set aside for verification and approval by the welding supervisor or for other action in the case of weld failures. Studs on approved test samples which can be used in production may have the bent studs straightened by a hollow pipe or hollow device placed over the stud and a continuous, slow load applied until the stud is in the correct position, or may be used bent at the discretion of the precast engineer.

Production welded studs requiring a bend shall be bent to the required position for embedment in a similar manner with a pipe or hollow device. The studs shall be bent with a slow, continuously applied load in such a manner that the bend is not made directly at the stud base but above the stud base so that a bend radius four (4) to six (6) times the stud diameter is made during the bending process. Studs shall not be

Commentary

Completion of the ten (10) stud tests successfully qualifies the process and the operator for production welding in the out-of-position or other application detail tested.

Standard

heated prior to or during bending operations.

During stud welding production, the operator shall ensure that all studs have the ceramic arc shield removed and are visually inspected at one hour intervals. This shall be considered a production lot of studs. The operator shall then proceed to inspect the production lot of studs.

If a visual inspection reveals any stud that does not show a full 360 deg. weld flash, the stud shall be measured to determine if the after weld length is within the satisfactory weld length reduction listed in Table 6.2.3a. Studs with satisfactory length reduction but lacking a full 360 deg. flash shall then be bent approximately 15 deg. from the original axis in a direction opposite to the missing portion of the weld flash. After bending, the stud shall be straightened by applying a slow, continuous load with a pipe or other hollow bending device. If no failure occurs, the stud weld shall be considered satisfactory. Threaded studs which exhibit lack of full 360 deg. weld flash shall be torque tested to the required proof load.

At the option of the welding supervisor, studs with a satisfactory after weld length but lacking a full 360 deg. weld flash may be repaired by adding a minimum fillet weld as required in Table 6.2.3b. The repair weld shall extend at least 3/8 in. (10 mm) beyond each end of the discontinuities being repaired.

If a repair weld is made on studs with lack of full 360 deg. weld flash and satisfactory after weld height, each stud repaired shall be subject to the 15 deg. bend and straighten test described above.

Studs with unsatisfactory weld burn off and lack of a full circumferential flash shall be repaired with a fillet weld, bent 15 deg. and straightened or torque tested.

Table 6.2.3b. Minimum fillet weld size for studs

Stud Diameter, in. (mm)	Minimum Size Fillet*, in. (mm)
1/4 (6) through 7/16 (11)	3/16 (5)
1/2 (12)	1/4 (6)
5/8, 3/4, 7/8 (16, 20, 22)	5/16 (8)
1 (25)	3/8 (10)

* Welding shall be done with low hydrogen electrodes 5/32 in. (4 mm) or 3/16 in. (5 mm) in diameter except that a smaller diameter electrode may be used on studs 7/16 in. (11 mm) diameter or under or for out-of-position welds.

Commentary

The weld flash around the stud base is inspected for consistency and uniformity. Lack of a flash may indicate a faulty weld. Fig C6.2.3a shows typical acceptable and unacceptable weld flash appearances. In Fig. C6.2.3a, (A) shows satisfactory stud weld with a good weld flash formation. In contrast, (B) shows a stud weld in which the plunge was too short. Prior to welding, the stud should always project the proper length beyond the bottom of the ferrule. (This type of defect may also be caused by arc blow.) (C) illustrates "hang-up." The stud did not plunge into the weld pool. This condition may be corrected by realigning the gun accessories to ensure completely free movement of the stud during lift and plunge. Arc length may also require adjustment. (D) shows poor vertical alignment, which may be corrected by positioning the stud gun perpendicular to the work. (E) shows the results of low weld power (heat). To correct this problem, the ground and all connections should be checked. Also, the current setting or the time setting, or both, should be increased. It may also be necessary to adjust the arc length. The effect of too much weld power (heat) is shown in (F). Decreasing the current setting or the dwell time, or both, will lower the weld power.

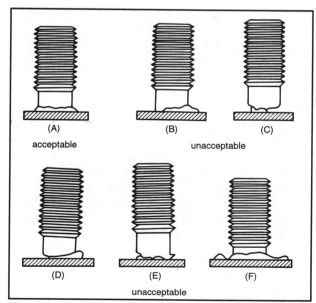

Fig. C6.2.3a. Satisfactory and unsatisfactory arc stud welds.

Standard

Studs which fail the inspection test by breaking off or tearing in the weld zone shall be removed. The weld spot shall be ground smooth and the stud replaced.

The welding supervisor shall verify that these inspections have been made and identify any studs repaired or replaced.

Inspection and testing during production welding for studs welded in positions other than down hand shall be identical to those procedures listed above. In addition, because of the greater risk of incomplete welding, these studs except for threaded studs shall receive 100% testing by striking with a heavy, short-handled machinists hammer. Any studs that break off shall be shown to the supervisor, and replaced with new studs or repaired by arc welding. Repaired or replaced studs shall be inspected by bending 15 deg. and straightening without failure.

After a production welding period of one hour, the two (2) stud pre-production testing procedure outlined above shall be repeated for studs welded downhand or in any other position. If either of the two (2) studs on this production testing fails the test, the supervisor and the operator together shall proceed as follows:

a. Working backward from the studs welded for this test, test the previous ten (10) most recently welded studs. If any fail, test ten (10) more and continue working backward until 10 studs are tested successfully without failure.

b. In addition, the supervisor shall visually inspect all studs welded from the ten (10) successful stud tests backward to the previous regular production test.

c. The pre-production test shall be repeated with appropriate equipment and weld setting adjustments until satisfactory welds are achieved before proceeding with further production.

The stud welding operator shall maintain a record of the start-up and hourly testing program testing and record the date, time of the test, conformance or non-conformance of the results and if non-conformance, the further testing required and performed. The sheet shall be kept in the stud welding operator's work area and shall be inspected at least daily by quality control inspector as a monitoring technique. In addition, the quality control inspector shall per-

Commentary

Standard

form bend tests and visual inspection on one stud for every 50 assemblies.

6.2.4 Special Testing

1. **Heat of Hydration.** When massive castings, or castings in partly insulated molds are being made, excessive concrete temperatures may be produced as the cement hydrates. This may, in turn, cause warpage or excessive shrinkage of the casting. When such castings are being produced, the temperature history of one typical casting shall be recorded by self-recording thermometers and, if over 150 deg. F (66 deg. C), steps must be taken in all castings to cool the concrete.

2. **Freeze-Thaw Tests.** Routine freeze-thaw durability tests of vertical elements are not required. If, due to special environmental conditions, freeze-thaw tests are specified, the test conditions of freezing in air, thawing in water (ASTM C666, Method B) shall be followed. Under this test method, the minimum allowable durability factor shall be 70.

6.3 Records

6.3.1 Recordkeeping

In order to establish evidence of proper manufacture and conformance with plant standards and project specifications, a system of recordkeeping shall be used which will provide full information regarding testing of materials, mix designs, production tests, and any other information specified for each project.

Commentary

C6.2.4 Special Testing

1. **Heat of Hydration.** In precast concrete members most of the heat generated by the hydrating cement is dissipated almost as fast as it is generated, and there is little temperature differential from the inside to the outside of the member.

   Extreme differences between internal and outside temperatures of massive members may result in surface cracking. For example, temperature stresses occur as the temperature of the concrete rises because of heat of hydration and then drops essentially to the temperature of its surroundings. As the outer surface cools and tends to shrink, compressive stresses are set up in the center and tensile stresses in the cooler outer surfaces. When these tensile stresses become greater than the tensile strength of the concrete, cracking occurs. Since the interior reacts so much more slowly than the surface to cycles of temperature, it is as though the surface were restrained by the interior concrete.

   The maximum permissible gradient for a nominally reinforced structure is generally less than 100 deg. F (38 deg. C).

2. **Freeze-Thaw Tests.** Laboratory freezing and thawing tests may be conducted to evaluate the durability of concrete under severe climatic conditions. These tests can be made on prismatic samples prepared from laboratory trial mixes or even from cores cut from the face of finished production units. Such tests, however, take several months to complete. The verticality of wall units seldom allows concrete to reach the critical saturation point (above 90 percent) on which such tests are based. However, where horizontal areas allow water or snow to accumulate or where ground level panels may be subjected to splashing by deicing salts and to freezing conditions at moisture contents above critical saturation, an air-entraining admixture should be used. In addition, it is probably a prudent policy to have air entrained concrete in all precast concrete exposed to freeze-thaw cycles.

   These tests take several months to complete. Equivalent evaluations can be obtained more rapidly by conducting "air void studies" (amount and character of entrained air in cores taken from the production unit) in accordance with ASTM C457.

C6.3 Records

C6.3.1 Recordkeeping

Records provide the vehicle for transmitting and evaluating, and should result in correction of deficient procedures or products by communication of problems.

Standard

Records shall be designed for a minimum of writing (see Sample Record Forms in Appendix E).

Each precast concrete unit shall be marked on the backside or edges of the unit with the date produced and a unique identification number that can be referenced to production and erection drawings and testing records.

Recordkeeping shall be the responsibility of the quality control inspection personnel. In the absence of project specification requirements or state statue, records shall be kept for a minimum of 5 years after final acceptance of the structure, or for the period of product warranty provided by the manufacturer, whichever is longer.

6.3.2 Suppliers' Test Reports

Certified test reports for materials not in-plant tested shall be required of suppliers. Refer to Article 2.1.2 for this requirement. These reports shall show the results of suppliers' mill or plant tests, tests by an independent testing laboratory, petrographic analysis of aggregates for face mixes, and other testing required by the project specifications. These reports shall state compliance with applicable specifications.

Mill or suppliers' test certificates or standard test results shall be available for the following materials:

1. Cement.
2. Aggregates.
3. Admixtures.
4. Reinforcing steel (all grades).
5. Prestressing tendons.
6. Studs or deformed anchors.
7. Structural steel or other hardware items.
8. Inserts or proprietary items as specified for individual projects.
9. Pigments.
10. Curing compounds.

These records shall be kept for the same period of time as the other project records.

Commentary

Units should be marked in an area of the member not exposed to view when in place on the structure, but accessible for review and inspection while in storage at the precast concrete plant.

Member markings, related to mold number when several identical molds are needed, make it possible to relate quality control data to specific units. They also facilitate documentation of member production, inspection and repair.

C6.3.2 Suppliers' Test Reports

All project specifications require standards for materials used in production. A manufacturing facility must obtain certificates of compliance to ensure the integrity of the product and to protect its own interests. Correlation to a specific project is needed for conformance with the design and specifications.

Plants which utilize materials that are fabricated or supplied by outside vendors should periodically inspect those vendors to assure that they are in compliance with the specifications of this manual. All vendors should be required to submit proof of compliance for both materials and workmanship.

Standard

6.3.3 Tensioning Records

An accurate record of all tensioning operations shall be kept and reviewed by qualified personnel. This record shall include, but not be limited to, the following:*

1. Date of tensioning.
2. Casting bed identification.
3. Description, identification and number of elements.
4. Manufacturer, size, grade, and type of strand.
5. Coil or pack number of strand, identifying heat.
6. Sequence of stressing (and detensioning, if critical).
7. Identification of jacking equipment.
8. For all pretensioning:
   a. Required total force per strand.
   b. Initial force.
   c. Anticipated and actual gauge pressure for each strand or each group of strands stressed in one operation.
   d. Anticipated elongation for each different jacking force.

   Tensioning calculations shall show a summary of anticipated operational losses such as strand chuck seating, splice chuck seating, abutment movements, thermal effects and self-stressing form shortening.
9. For single strand pretensioning:
   a. If the setup is open, where no friction is imposed on the strand between strand chucks, the record shall contain jacking force for each strand and the actual elongation of the first strand tensioned in each different stress group and the elongation of at least 10% of the remaining strand.
   b. If one system is used for determination of the jacking load and friction is expected to be imposed on the strand, the actual elongation of every strand and the jacking force shall be recorded.
10. For multiple strand pretensioning, the actual elongation of the strand group measured by movement of the jacking header.
11. For post-tensioning:
    a. The actual net elongation of each tendon

---
*Sample record form is shown in Appendix E.

Commentary

C6.3.3 Tensioning Records

Variations of actual values from computed theoretical values should be computed each day to monitor developing trends and to make personnel aware of tolerances.

Standard

with allowance made for elastic shortening of the member, as well as jacking force.

   b. Data on and date of grouting.

12. Any unanticipated problems encountered during tensioning such as wire breakage, excessive seating, restressing, or other factors having an influence on the net stress.

6.3.4 Concrete Records

Records of concrete operations and tests shall be kept so the following data will be available:*

1. Unit and job identification.
2. Production date.
3. Mix proportions by weight.
4. Mixing water corrections and/or aggregate corrections due to surface moisture.
5. Yardage, design and actual yield or unit weight.
6. Identification of production area, mold or bed.
7. Test specimen identification.
8. Concrete temperature.
9. Air temperature, weather conditions, if applicable, and any measures taken for cold or hot weather concreting.
10. Slump.
11. Air content.
12. Unit weight (fresh).
13. Inspection of batching, mixing, conveying, placement, consolidation and finishing of concrete.
14. Method and duration of curing, e.g., temperature charts for accelerated curing.
15. Strength at stress release or stripping.
16. 28-day strength.
17. Absorption on concrete exposed to weathering.
18. Cylinder strength tests and air-dry unit weight for lightweight concrete.
19. Fresh unit weight for lightweight concrete.
20. Inspection reports.

6.3.5 Calibration Records for Equipment

Calibration records for plant equipment such as batch plant scales, compression testing machine,

Commentary

C6.3.4 Concrete Records

In evaluating mix design efficiency or performance, all information as listed is needed to eliminate variables. If a problem occurs, information is needed for evaluation in the same manner.

C6.3.5 Calibration Records for Equipment

Calibration records are required as specified for each type of equipment and equipment should be re-calibrated as required.

Standard

impact hammer, or non-destructive testing devices and other necessary equipment shall be supplied by the testing agency and the equipment operator shall have ready access to the records.

6.4 Laboratory Facilities

6.4.1 General

The plant shall maintain an adequately equipped laboratory or retain the services of a testing agency in which investigation and development of suitable concrete mixes may be conducted, and ongoing quality control testing may be performed.

The laboratory facilities shall be in a protected area with environmental controls to ensure proper working conditions. Laboratory equipment shall be maintained in proper condition and calibrated as needed, but not less than annually. Calibration records shall be kept on file.

6.4.2 Quality Control Testing Equipment

The plant shall have all equipment required for performing the testing procedures. Equipment shall meet the requirements of the test procedure specification.

6.4.3 Test Equipment Operating Instructions

Operating instructions shall be obtained for all testing equipment as well as national and industry standards, for materials and testing. These instructions shall be kept in the laboratory and shall be carefully followed by all testing personnel.

Testing machines shall be kept clean and no attempt shall be made to use them beyond their rated capacities. Machines shall be capable of applying loads at the specified rate. Testing machines shall be calibrated so that the maximum error is not more than ±1 percent of full scale reading. Calibration shall be performed whenever there is reason to question the

Commentary

Records that show deviations should be used by plant personnel to obtain correct readings.

C6.4.2 Quality Control Testing Equipment

The laboratory should have the facilities necessary for the development and assessment of concrete mixes and the quality control tests to be performed by the manufacturer in accordance with ASTM C1077.

Since the compressive strength of concrete must be determined prior to stripping or stress transfer, testing equipment should be provided at all plants. It is preferable to have this equipment at the plant to avoid delays and possible damage to cylinders by transporting them to a central laboratory.

Testing machines should be equipped with a guard that will protect personnel from flying debris.

Care in using capping compounds is required because of the potential for fire. If the compound is overheated, a fire may result, so a functioning thermostat is required. Due to fumes produced, the capping compound heater should be vented to the exterior of the testing area.

Standard

accuracy of indicated loads, or at least annually. Calibration curves shall be available at all times and used by testing personnel.

Commentary

DIVISION 7 — PRODUCT TOLERANCES

Standard

7.1 Requirements for Finished Product

7.1.1 Product Tolerances – General

The tolerances listed in this Division shall govern unless other tolerances are noted in the contract documents for a specific project.

Commentary

C7.1 Requirements for Finished Product

C7.1.1 Product Tolerances — General

Tolerances are divided into three categories: product tolerances, erection tolerances and interfacing tolerances. See Appendix I for erection tolerances.*

Tolerance is a specified permissible variation from exact requirements of the contract documents. A tolerance may be expressed as an additive or subtractive (±) variation from a specified dimension or relation or as an absolute deviation from a specified relation. Tolerances should be established for the following reasons:

1. **Structural.** To ensure that the structural design properly accounts for factors sensitive to variations in dimensional control. Examples include eccentric loading conditions, bearing areas, reinforcement locations, and hardware and hardware anchorage locations.

2. **Feasibility.** To ensure acceptable performance of joints and interfacing materials in the finished structure and to ensure that designs and details are dimensionally feasible from manufacturing and construction points of view.

3. **Visual effects.** To ensure that the variations will be controllable and result in an acceptable looking structure.

4. **Economics.** To ensure ease and speed of production and erection by having a known degree of accuracy in the dimensions of the precast concrete units.

5. **Legal.** To avoid encroaching on property lines and to establish a tolerance standard against which the work can be compared in the event of a dispute.

6. **Contractual.** To establish a known acceptability range and also to establish responsibility for developing, achieving and maintaining mutually agreed tolerances.

It is very important that the entity (architect, engineer of record, or precaster) taking responsibility for establishing the project tolerances be clearly defined at the onset of the project.

The architect/engineer should be responsible for coordinating the tolerances for precast concrete work with the requirements of other trades whose work adjoins the precast concrete construction. In all cases the tolerances must be reasonable, realistic and within generally accepted limits. It should be understood by those involved in the design and construction process that tolerances shown in Article 7.1.2 must be considered as guidelines for an acceptable range and not limits for rejection. If these tolerances are met, the unit should be accepted. If

---

* See the full report of the PCI Committee on Tolerances, "Tolerances for Precast and Prestressed Concrete," PCI JOURNAL, V. 30, No. 1, January-February, 1985, for a complete discussion on erection and interfacing tolerances.

Standard

Applicable product tolerances shall be clearly conveyed to production and quality control personnel.

Accurate measuring devices and methods with precision capability appropriate to the tolerance being controlled shall be used for both setting and checking product tolerances. To maximize accuracy, products shall not be measured in increments in a manner which creates the possibility of cumulative error.

Any special measuring or record keeping methods specified in the contract documents shall be observed by the plant quality control personnel.

7.1.2 Product Tolerances

Commentary

these tolerances are exceeded, the unit may still be acceptable if it meets any of the following criteria:

1. Exceeding the tolerance does not affect the structural integrity or architectural performance of the unit.

2. The unit can be brought within tolerance by structurally and architecturally satisfactory means.

3. The total erected assembly can be modified to meet all structural and architectural requirements.

Where a project involves particular features sensitive to the cumulative effect of generally accepted tolerances on individual portions, the architect/engineer should anticipate and provide for this effect by setting a cumulative tolerance or by providing escape areas (clearances) where accumulated tolerances can be absorbed. The consequences of all tolerances permitted on a particular project should be investigated to determine whether a change is necessary in the design or in the tolerances applicable to the project or individual components.

Careful inspection of the listed tolerances will reveal that many times one tolerance will override another. The allowable variation for one unit of the structure should not be applicable when it will permit another unit of the structure to exceed its allowable variations. Restrictive tolerances should be reviewed to ascertain that they are compatible and that the restrictions can be met. For example, a requirement which states that "no bowing, warpage or movement is permitted" is not practical or possible to achieve.

C7.1.2 Product Tolerances

Product tolerances are those needed in any manufacturing process. They are normally determined by economical and practical production considerations, and functional and appearance requirements.

The architect/engineer should specify product tolerances or require performance within generally accepted limits. Tolerances for manufacturing are standardized throughout the industry and should not be made more exacting, and therefore more costly, unless absolutely necessary.

Standard

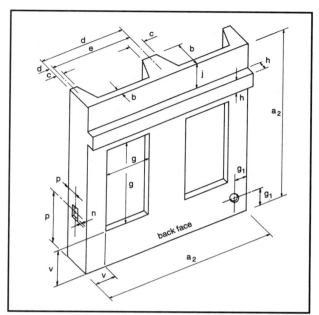

Fig. 7.1.2a

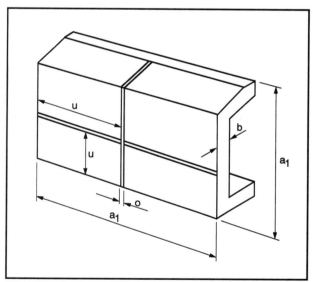

Fig. 7.1.2b

During the pre- and post-pour check of dimensions, the inspector shall have the approved shop drawings for reference. Discrepancies shall be noted on the post-pour record and transmitted to management or engineering for their evaluation.

Figs. 7.1.2a and b shows the location of the tolerances listed below.

Commentary

Standard

Units* shall be manufactured so that the face of each unit which is exposed to view after erection complies with the following dimensional requirements:

a_1 Overall height and width of units measured at the face exposed to view.

    10 ft or under .... ± 1/8 in. (±3 mm)

    10 to 20 ft ...+ 1/8 in., –3/16 in. (+3 mm, –5 mm)

    20 to 40 ft .... ± 1/4 in. (±6 mm)

    Each additional 10 ft .. ± 1/16 in. (±1.5 mm) per 10 ft (3 m)

a_2** Overall height and width of unit measured at the face not exposed to view.

    10 ft or under .... ± 1/4 in.

    10 to 20 ft .... + 1/4 in., –3/8 in.

    20 to 40 ft .... ± 3/8 in.

    Each additional 10 ft .... ± 1/8 in.

b Total thickness or flange thickness .... + 1/4 in., –1/8 in. (+6 mm, –3 mm)

c Rib thickness .... ± 1/8 in. (±3 mm)

d Rib to edge of flange .... ± 1/8 in. (±3 mm)

e Distance between ribs .... ± 1/8 in. (±3 mm)

f Variation from square or designated skew (difference in length of the two diagonal measurements) .... ± 1/8 in. (±3 mm per 2 m) per 6 ft of diagonal or ± 1/2 in. (±13 mm), whichever is greater.***

g Length and width of blockouts and openings within one unit .... ± 1/4 in. (±6 mm)

g_1 Location and dimensions of blockouts hidden from view and used for HVAC and utility penetrations .... ± 3/4 in. (±19 mm)

h Dimensions of haunches .... ± 1/4 in. (±6 mm)

i Haunch bearing surface deviation from specified plane .... ± 1/8 in. (±3 mm)

---

* For non-architectural precast concrete units, such as columns, beams, etc., tolerance requirements are given in PCI MNL-116, *Manual for Quality Control for Plants and Production of Precast Prestressed Concrete Products.*

** Unless joint width and fit up requirements require more stringent tolerances.

*** Applies both to panel and to major openings in the panel.

Commentary

a_1. Length or width dimensions and straightness of a unit will affect the joint dimensions, opening dimensions between panels, and perhaps the overall length of the structure.

b. Thickness variation of the precast concrete unit becomes critical when interior surfaces are exposed to view. A non-uniform thickness of adjacent panels will cause offsets of the front or the rear faces of the panels.

f. Panels out-of-square can cause tapered joints and make adjustment of adjacent panels difficult.

Standard

j   Difference in relative position of adjacent haunch bearing surfaces from specified relative position .... ± 1/4 in. (±6 mm)

k   Bowing .... ± L/360
      max. 1 in. (25 mm)

l   Local smoothness .... 1/4 in. in 10 ft (6 mm in 3 m) Does not apply to visually concealed surfaces. (Refer to Fig. C7.1.2c for definition.) For clay

Commentary

k. Bowing is an overall out-of-planeness condition which differs from warping in that while the corners of the panel may fall in the same plane, the portion of the panel between two parallel edges is out of plane. Bowing conditions are shown in Fig. C7.1.2a.

Bowing and warping tolerances have an important effect on the edge match up during erection and on the visual appearance of the erected panels, both individually and when viewed together. The requirements for bowing and warping of panels may be overridden by tolerances for panels as installed with reference to joint widths, jog in alignment and step in face.

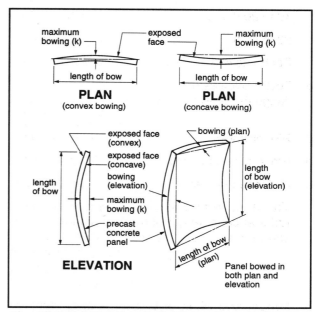

Fig. C7.1.2a. Bowing definitions for panels

A special appearance requirement may be necessary for honed or polished flat concrete walls where bowing or warping tolerances might have to be decreased to 75 or 50% of tolerances in Article 7.1.2 in order to avoid joint shadows. Another special case might be tolerances for dimensions controlling the matching of open shaped panels. These tolerances may have to be smaller than the standard dimensional tolerance (75% or 50%), unless the architect has recognized and solved the alignment problem as part of the design.

l. Surface out-of-planeness, which is not a characteristic of the entire panel shape, is defined as a local smoothness variation rather than a bowing variation. Examples of local smoothness

Standard

product faced units 1/16 in. (2 mm) variation between adjacent clay products.

Commentary

variations are shown in Fig. C7.1.2b. The tolerance for this type of variation is usually expressed in fractions of an inch per 10 ft (mm per 3 m).

Fig. C7.1.2b also shows how to determine if a surface meets a tolerance of 1/4 in. in 10 ft (6 mm per 3 m). A 1/4 in. (6 mm) diameter by 2 in. (50 mm) long roller should fit anywhere between the 10 ft (3 m) long straightedge and the element surface being measured when the straightedge is supported at its ends on 3/8 in. (10 mm) shims as shown. A 1/2 in. (12 mm) diameter by 2 in. (50 mm) long roller should not fit between the surface and the straightedge.

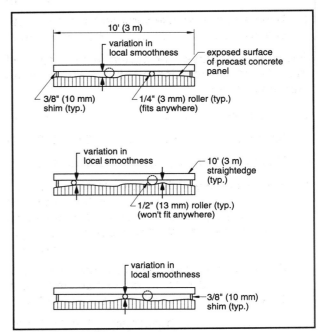

Fig. C7.1.2b. Measuring local smoothness variations.

---

m Warping .... 1/16 in. per ft (1.5 mm per 300 mm) of distance from nearest adjacent corner

m. Warping is generally an overall variation from planeness in which the corners of the panel do not all fall within the same plane. Warping tolerances are stated in terms of the magnitude of the corner variation, as shown in Fig. C7.1.2c.

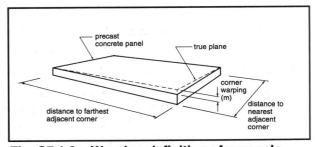

Fig. C7.1.2c. Warping definitions for panels

Standard

n Tipping and flushness of plates .... ± 1/4 in. (±6 mm)

o Dimensions of architectural features and rustications .... ± 1/8 in. (±3 mm)

Positions tolerances. For cast-in items measured from datum line location as shown on approved erection drawings:

p Weld plates .... ± 1 in. (±25 mm)

q Inserts .... ± 1/2 in. (±13 mm)

r Handling devices .... ± 3 in. (±75 mm)

s_1 Reinforcing steel and welded wire fabric .... ± 1/4 in. (±6 mm) where position has structural implications or affects concrete cover, otherwise ± 1/2 in. (±13 mm)

s_2 Reinforcing steel extending out of member .... ± 1/2 in. (±13 mm) of plan dimensions

t Tendons ....vertical: ± 1/4 in. (±3 mm)
....................horizontal: ± 1 in. (±25 mm)

u Location of rustication joints ± 1/8 in. (±3 mm)

v Location of opening within panel ...... ± 1/4 in. (±6 mm)

w Flashing reglets .... ± 1/4 in. (±6 mm)

x Flashing reglets at edge of panel ..... ± 1/8 in. (±3 mm)

y Reglets for glazing gaskets .... ± 1/8 in. (±1.5 mm)

z Electrical outlets, hose bibs, etc. ...... ± 1/2 in. (±13 mm)

aa Haunches .... ± 1/4 in. (±6 mm)

bb Allowable rotation of plate, channel insert, electrical box, etc., Fig. 7.1.2c .... 2 deg. rotation or 1/4 in. (6 mm) max. over the full dimension of the element

Commentary

Note that bowing and warping tolerances are of primary interest at the time the panel is erected. Careful attention to pre-erection storage of panels is necessary, since storage conditions can be an important factor in achieving and maintaining panel bowing and warping tolerances.

The likelihood that a panel will bow or warp depends on the design of the panel and its relative stiffness or ability to resist deflection as a plate member. Slender panels are more likely to bow, and the tolerances should be more liberal.

Standard

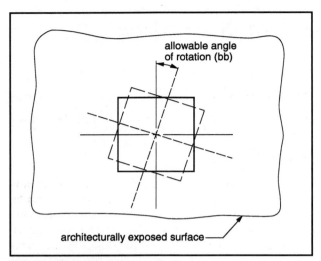

Fig. 7.1.2c. Allowable rotation.

Commentary

APPENDIX A

Guidelines for Developing Plant Quality System Manual

INTRODUCTION

Documented quality system procedures should be the basis for the overall planning and administration of activities which impact on quality. These documented procedures should cover all the elements of the company's quality system standard. They should describe (to the degree of detail required for adequate control of the activities concerned) the responsibilities, authorities, and interrelationships of the personnel who manage, perform, verify, or review work affecting quality, how the different activities are to be performed, the documentation to be used, and the controls to be applied. All of this is particularly important for personnel who need the organizational freedom and authority to (a) initiate action to prevent non-conformances of any kind; (b) identify and document any problems relating to product, process and/or quality systems; (c) initiate, recommend or provide solutions through designated channels; (d) verify the implementation of solutions; and (e) control further processing, delivery or installation of non-conforming production until the problem(s) have been corrected.

Documented quality system procedures should not, as a rule, enter into purely technical details of the type normally documented in detailed work instructions.

The quality manual should identify the management functions, address or reference the documented quality system and procedures, and briefly cover all the applicable requirements of the quality system standard selected by the organization. Wherever appropriate, and to avoid unnecessary duplication, reference to existing recognized standards or documents available to the quality manual user should be incorporated.

Release of the quality manual should be approved by the management responsible for its implementation. Each copy should bear evidence of this release authorization.

Although there is no required structure or format for a quality manual, it should convey accurately, completely, and concisely the quality policy, objectives, and governing documented procedures of the organization.

One of the methods of assuring that the subject matter is adequately addressed and located would be to key the sections of the quality manual to the quality elements of the PCI quality manual (MNL-117). Other approaches, such as structuring the manual to reflect the nature of the organization, are equally acceptable.

MANUAL CONTENTS

A quality manual should normally contain the following:

Table of Contents

The table of contents of a quality manual should show the titles of the sections within it and how they can be found. The numbering or coding system of sections, subsections, pages, figures, exhibits, diagrams, tables, etc., should be clear and logical and should include revision status.

Definitions

Definitions of terms or concepts that are uniquely used within the plant should be included, although it is recommended, when practical, to use standard definitions and terms shown in MNL-117.

   I. **MANAGEMENT RESPONSIBILITY**

      A. **Quality Policy Statement**

         1. In the most general sense, a quality policy should be a short, clear statement of commitment to a standard of quality and a warranty from the highest level of management that quality will not be compromised when it is in conflict with other immediate interests. The quality policy also should define objectives pertaining to quality. The objectives focus and direct the quality system toward concrete goals, giving the plant's personnel the motivation to develop and maintain the system. To the customer, the objectives are an expression of an implied promise that satisfaction of their needs is going to be the point of reference in their relationship with the plant.

         2. This section should also describe how the quality policy is made known to, and understood by, all employees and how it is implemented and maintained at all levels. Management should ensure that individuals are familiar with those contents of the manual appropriate to each user within the organization.

         Include a brief description of the documented procedures used to identify the status and to control the distribution of the quality manual, whether or not it contains confidential information, whether it is used only for the plant's internal purposes, or whether it can be made available externally.

B. Organization

1. Responsibility and Authority

    a. This section of the manual should provide a graphical organization chart showing key personnel, their duties and responsibilities, authorities and the interrelationship structure. This is the most effective and straightforward way to define and document an organization's structure. Include a general organization chart for the whole company, supplementing it with more detailed charts that present internal organizations of departments directly concerned with the QA and QC activities. It is not practical nor required to include in the charts the names of assigned personnel. Documents evidencing individual assignments to organizational functions shall be maintained elsewhere - for example, in the personnel department.

    b. While the charts document general functional responsibilities, there is also a need to assign personnel the authority and responsibility to carry out specific actions referenced in the quality system. Details of the responsibilities, authorities, and hierarchy of all functions which manage, perform, and verify work affecting quality should be provided. Assignment and documentation of those specific responsibilities is best made directly within procedures dealing with the corresponding actions. For example, personnel responsible for identifying and recording product quality problems can be defined in the inspection and testing procedures section.

2. Verification Resources and Personnel

    a. Procedures and quality plans should completely define the review, monitoring, inspection and testing needs at specific points in purchasing, receiving, manufacturing and shipping. The extent and scope of the verifications must be established by the plant.

    The verification activities should be supported with qualified personnel and adequate resources. The plant should identify what level of training and experience is needed to perform specific verification functions and to evidence that the assigned personnel meet those requirements. All training, no matter how informal, should be documented and recorded.

    b. Self-inspection may be adopted provided that it is qualified, documented and regularly audited. Audits should be carried out by personnel independent from those having direct responsibility for the work. Inspection and testing are excluded from this requirement of independence.

C. Management Review

In addition to analysis of PCI Plant Certification audit results, rules for scheduling, conducting and recording management reviews of the quality system should be established.

II. QUALITY SYSTEM

The manual should describe and document the applicable elements of the plant quality management system. The description should be divided into logical sections revealing a well-coordinated quality system. The quality manual should include policies, operating procedures, work instructions, process procedures, company standards, PCI standards, and the production and quality plans. This may be done by inclusion of, or reference to, documented quality system procedures. Auditing and review of the implementation of the quality system should be discussed.

The purpose is to:

1. Define purpose, contents and format of the quality system documentation.

2. Assign responsibilities for establishing and maintaining the quality system documentation.

III. DOCUMENT CONTROL

The purpose, scope and responsibility for controlling each type of quality system document should be defined. This section should provide a system and instructions and assign responsibilities for establishment, review, authorization, issue, distribution and revisions of the quality system documents.

A brief description of how the quality manual is revised and maintained, who reviews its contents and how often, who is authorized to change the quality manual, and who is authorized to approve it; this information may also be given under the system element concerned; a method for determining the history of any change in procedure may be included, if appropriate.

To ensure that each manual is kept up to date, a method is needed to assure that all changes are received by each manual holder and incorporated into each manual. A table of contents, a separate revision-status page, or other suitable means should be used to assure the users that they have the authorized manual.

IV. PURCHASING

There should be a clear and full description of ordered products and the vendor monitoring procedures to verify that quality requirements of the plant are met. Procedures for disposition of nonconforming materials should be described.

Rules applicable to preparation, review and approval of purchasing documents and use of approved vendors should be provided.

V. PRODUCT IDENTIFICATION AND TRACEABILITY

Describe the system to readily identify each unit produced and to distinguish between different grades of otherwise similar materials, components, subassemblies and products and maintenance procedures for records.

VI. PROCESS CONTROL

Process covers all activities connected with production such as production planning, environment, equipment, technology, process control, work instructions, product characteristics control, criteria for workmanship, and so forth.

The production plan should define, document and communicate all manufacturing processes and inspection points as well as workmanship standards. A production flow chart should be included. Work instructions should indicate how to operate and adjust equipment, what steps are required to perform certain operations and inspections, warn against safety hazards, etc. Maintenance and calibration of equipment and testing apparatus schedules should be established and recorded.

Provide for a system and instructions, and assign responsibilities for:

1. Establishing and use of work order, work instructions and change orders.

2. Production equipment checking and monitoring.

3. Qualification and control of special processes such as welding.

4. Establish criteria and responsibility for maintenance of the production environment - items that adversely affect performance.

VII. INSPECTION AND TESTING

A. Receiving Inspection and Testing

The purpose of this section is to provide for a system and instructions, and to assign responsibilities for performing and recording the receiving inspections of purchased products. Scope and form of receiving inspections should be established.

As a minimum, the scope of receiving QC inspections comprises:

1. Review of material certification, source inspection and test records, compliance certificates and other such documentation delivered with the product.

2. Visual inspection to detect any damage or other visible quality problems.

3. Taking measurements and testing, as required.

4. Recording the actual measurements and test results.

B. In-Process Inspection and Testing

Degree, scope and manner of in-process inspections should be established to ensure products are produced in accordance with production drawings and approved samples. Inspection of special processes such as welding must be monitored and controlled. Include sample copies of checklists or forms used by plant personnel for quality control functions.

Specific requirements to be covered are:

1. Planning and documentation of inspection in the company's quality plans or procedures.

2. Handling of changes to shop drawings during production.

3. Identification of inspection status of product.

4. Handling of non-conforming product.

Identify those product characteristics that can be inspected only at specific stages of production.

C. Final Inspection

The quality plan and procedures should define the extent and scope of the final inspections and tests to verify that all receiving and in-process inspections specified for the product have been carried out with satisfactory results. The means of identifying non-conforming products should be described.

D. Inspection and Test Records

Describe the recording of each inspection, sign-off procedure and the maintenance of the records.

VIII. INSPECTION, MEASURING AND TEST EQUIPMENT

The purpose of this section is to provide for a system and instructions, and to assign responsibilities for calibration at prescribed intervals, identification including type, model, range and accuracy, and maintenance of measuring and test equipment. Calibration procedures and records should be established and maintained.

Identification of measurements to be made and the tolerances allowed should be documented in the quality plan, product drawings and specifications for use by production and inspection personnel.

The system of checking and certifying jigs, templates and patterns or molds used in manufacturing or inspection should be established.

IX. INSPECTION AND TEST STATUS

The purpose of this section is to provide for a system and instructions, and to assign responsibilities for:

1. Identification of a product's inspection status.
2. Release of conforming product for storage or shipment.

The most common methods for inspection status identification are marking, tagging or labeling the product itself, entering an inspection record in a traveler accompanying the product, and physical segregation of products according to their inspection status. Whatever system is adopted, it must be clearly documented in appropriate procedures.

X. CONTROL OF NON-CONFORMING PRODUCT

The purpose of this section is to provide for a system and instructions, and to assign responsibilities for:

1. Identifying, documenting and evaluating non-conforming incoming materials, materials in production, products that preclude repair and products that are repairable.

2. Disposition of a non-conforming product.

Procedures in this section must demonstrate clearly that the non-conforming product is being prevented from shipping and that the non-conformity is recorded. Reinspection procedures for repaired or reworked products should be described.

A non-conformity may be an isolated minor defect that can be repaired immediately by a simple process or, at the other extreme, a serious public safety hazard potentially involving a large number of already shipped products. In dealing with this problem, it is useful to divide the possible non-conformities into categories such as minor or major or cosmetic or structural defects and assign to them different levels of authority for disposition. The classes should distinguish between small and large number of products affected, isolated errors and system problems, possibility of impact on already shipped products and seriousness of the non-conformity. Such classification will also be very helpful in planning and implementing corrective actions. The responsibility and procedure for the preparation of concrete repair mixes should be defined.

XI. CORRECTIVE ACTION

All activities relating to corrective actions should be covered by written procedures. The corrective action system should comprise investigation of causes, implementation of corrective actions and verification of their effectiveness.

XII. HANDLING, STORAGE, AND LOADING/DELIVERY

A. General

The purpose of this section is to define specific rules for handling different units, prescribe a management system for stored units, and specify arrangements for protection of products during transportation

B. Handling

Describe procedures to regulate the use and instruct in operating handling equipment as well as handling equipment maintenance.

C. Storage

The purpose of this section is to provide for a system and instructions, and to assign responsibilities for:

1. Ensuring products are stored in accordance with drawings.
2. Use and maintenance of storage areas for both materials and finished product.
3. Periodical assessment of stored materials and product to check on damage, stains or contamination.

D. Loading/Delivery

Describe procedure to provide for a system and instructions, and to assign responsibilities for loading products and protecting products during delivery whether or not delivery is required by contract.

XIII. QUALITY RECORDS

Compliance with the following requirements should be documented in written procedures:

1. Identifiable and legible records.
2. Easily retrievable records from files in a suitable environment.
3. Retention of records for a specified period of time.

Format, identification, applicable processing and filing location for a record should be stipulated in the procedure that requires creation of the record. There should be an index listing types of records and their locations.

XIV. INTERNAL QUALITY AUDITS

The purpose of this section is to provide for a system and instructions, and to assign responsibilities for conducting and documenting internal quality audits to verify compliance to the quality manual and the effectiveness of activities to achieve defined quality objectives. The audit plan should list all the activities in the various sections of the company's quality manual, identify the location where the activities are taking place, and schedule the audit for each activity and location. Activities which receive more frequent auditing by plant personnel should be described. A corrective action and follow-up procedure for any deficiencies found during an audit should be established and documented to verify the implementation and effectiveness of the corrective action. In addition, discuss the external audit procedures of the PCI Plant Certification program and the corrective action and follow-up procedures.

XV. TRAINING

The purpose of this section is to provide for a system and instructions, and to assign responsibilities for determining training needs, providing the training and keeping training records.

As a minimum, the training should comprise the following topics:

1. Product orientation with emphasis on crucial quality characteristics.
2. Presentation of the company's quality system.
3. The role of employees in maintaining the quality system and improving its efficiency.

Recording procedures for employee participation in training and maintenance of the records should be discussed.

Example of a possible format for a section of a quality manual

Organization	Title-Subject		Number	
Unit issuing	**Approved by**	**Date**	**Revision**	**Page**

POLICY OR POLICY REFERENCE

Give governing requirement.

PURPOSE AND SCOPE

List why, what for, area covered, and exclusions.

RESPONSIBILITY

Give organizational unit responsible for implementing the document and achieving the purpose.

ACTIONS AND METHODS TO ACHIEVE SYSTEM ELEMENT REQUIREMENT

List, step by step, what needs to be done. Use references, if appropriate. Keep in logical sequence. Mention any exceptions or specific areas of attention. Consider the use of flowcharts.

DOCUMENTATION AND REFERENCES

Identify which referenced documents or forms are associated with using the document, or what data have to be recorded. Use examples, if appropriate.

RECORDS

Identify which records are generated as a result of using the document, where these are retained, and for how long.

NOTES

1. This format may also be used for a documented quality system procedure.
2. The structure and order of the items listed above should be determined by organizational needs.
3. The approval and revision status should be identifiable.

Typical quality system document hierarchy

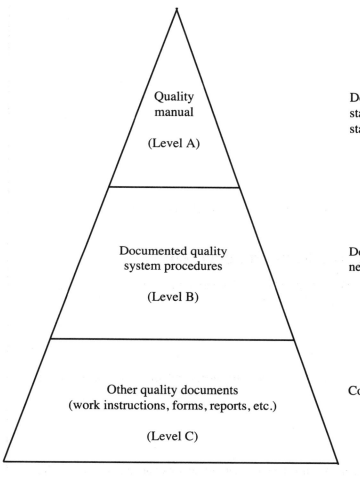

Document contents

Describes the quality system in accordance with the stated quality policy objectives and the applicable standard.

Describes the activities of individual functional units needed to implement the quality system elements.

Consists of detailed work documents.

NOTE: Any document level in this hierarchy may be separate, used with references, or combined.

APPENDIX B

Design Responsibilities and Considerations

General

Design calculations should be performed under the supervision of a registered professional engineer with experience in precast design and should be stamped (sealed). The precast manufacturer should be prepared to assist in the design of panels and connections. The owner's designer (architect/engineer) should provide all of the design load information and maintains overall design responsibility.

The architect/engineer inexperienced in precast design can benefit from early contact with experienced manufacturers who can offer constructive advice during preliminary design.

It is common practice for the architect/engineer to rely on the precast manufacturer for handling and erection procedures, and for ensuring that the unit is adequately designed for the loads incurred during manufacturing, handling, shipping, and installation. All procedures should be checked to ensure that they do not cause: (1) cracking, (2) structural damage, (3) architectural impairment, or (4) permanent distortion.

Contract drawings prepared by the architect/engineer should show connection locations in sufficient detail to permit design, estimating, and bidding. Panel manufacturers, during the preparation of shop drawings, should review connections for tolerances, clearances, practicality, and performance. The manufacturer should call to the architect/engineer's attention any foreseen problems. Whether all reinforcement and connection details are shown by the designer or left entirely to the precaster is the prerogative of the architect/engineer. This decision may be determined by local practices or by the project specifications.

The general contractor is responsible for the location of all panel bearing surfaces and anchorages on non-precast concrete structural frames. Changes, other than adjustments within the prescribed tolerances, require approval by the architect/engineer.

Architect/Engineer Responsibilities

Responsibility for both structural and aesthetic design of architectural precast concrete should rest with the architect/engineer who should:

1. Provide clear and concise drawings and specifications and, where necessary, interpretation of the contract documents. Identify the design loads and applicable codes on the contract drawings.

2. Establish the standards of acceptability for surface finish, color range and remedial procedures for defects and damage. The architect/engineer's right to reject work is provided to protect the owner against work which does not meet these standards.

3. Determine the part, if any, played by the precast concrete in the support of the structure as a whole.

4. As part of the design, allow for the effect of differences in material properties, stiffness, temperatures, and other elements which might influence the interaction of precast concrete units with the structure.

5. Determine the load reactions necessary for the accurate design of both reinforcement and connections.

6. Evaluate thermal movements as they might affect requirements for joints, connections, reinforcement and compatibility with adjacent materials.

7. Analyze the water-tightness of concrete wall panel systems, evaluating joint treatment, including the performance of adjacent materials for compatibility in joint treatment, and the proper sealing of windows and other openings, except where such systems are manufactured and marketed as proprietary items.

8. Make selection of surface finishes recognizing certain limitations in materials and production techniques in regard to uniform color, texture, repairability and performance, especially the limitations which are inherent in natural materials.

9. Design the supporting structure so that it will carry the weight of the precast concrete as well as any superimposed loads; including provisions for deflection and rotation of the supporting structure during and after erection of the precast concrete. Specify which members of the supporting structure are to support the architectural precast concrete, e.g., beams, columns or slabs.

10. Design the supporting structure for the temporary loading conditions associated with the proposed erec-

tion sequence and methods. However, the general contractor is responsible for the construction means, methods, techniques, sequences, or procedures.

11. Provide details for the interfacing of precast concrete and other materials and coordinate with general contractor.

12. Make selection of interior panel finishes defining the area of exposure and the interior appearance for occupancy requirements, again recognizing material and production limitations.

13. Design sandwich wall panels, where applicable, for proper structural and insulating performance.

14. Design for durable exterior walls with respect to weathering, corrosive environments, heat transfer, fire resistance rating, vapor diffusion, and moist air or rain penetration.

15. Review and approve precaster's shop drawings (erection and typical shape drawings) and typical calculations.

16. Specify dimensional and erection tolerances for the precast concrete, and tolerances for supporting structure and contractor's hardware. Any exceptions to PCI standard tolerances should be clearly identified.

Contract drawings should provide a clear interpretation of the configuration and dimensions of individual units, their relation to the structure, and to other materials. The drawings must supply the following information:

1. All sections and dimensions necessary to define the size and shape of the unit.

   The transition from one mass of concrete to another within a precast concrete unit is a prime consideration in finalizing shape. Wherever possible, this transition should be gradual. As an example, a large radius or fillet is preferred for the transition from a mullion, sill, or eyebrow to a flat portion within a precast concrete element. This is necessary to minimize the possibility of shrinkage stresses causing crack formation at the point where two concrete sections of distinctly different mass meet. A gradual flowing transition of shape along with well distributed reinforcement will reduce the risk of this type of cracking. Where a reasonable transition is not possible, a concrete mix with low shrinkage characteristics (low water and cement content) becomes increasingly important. Alternatively, the transition may be accomplished by two stage (sequential) casting.

2. Location of all joints, both real (functional) and false (aesthetic). Joints between units should be completely detailed.

3. The materials and finishes required on all surfaces, and a clear indication of which surfaces are to be exposed when in place.

   When the surface of a precast concrete element has two or more different mixes or finishes, a demarcation (reveal) feature is a necessary part of the design. The depth of the groove should be at least 1-1/2 times the aggregate size and the width should be in dimensional lumber increments such as 3/4 or 1-1/2 in. (19 to 38 mm). The groove should generally be wider than it is deep in order to strip the element without damaging the mold. The importance of the separation provided by a demarcation feature depends on the configuration of the unit on which the finishes are combined. For example, a groove or offset is necessary when an exposed aggregate flat surface is located between widely spaced ribs with a different surface finish, but not necessary when a similar flat surface lies between closely spaced ribs. Proper samples should be used to assess the problem. The importance of the separation also depends on the specific types of finishes involved. If a demarcation groove occurs near a change of section, it may create a weakness and counter any attempt to provide a gradual transition from one mass to another.

4. Details for the corners of the structure.

5. Details for jointing to other materials.

6. Details for unusual conditions and fire endurance requirements.

7. Design loads and moments, which include eccentricities.

8. Deflection limitations.

9. Specified tolerances and clearance requirements for proper product installation.

10. Support locations for gravity and lateral loads.

11. Specific mention of any required erection sequences.

12. Reinforcement of units, see Option 1, Table B1.

13. Details of connections to the supporting structure, see Option 1, Table B1.

    This information should be sufficiently detailed to enable the precaster to design and produce the units and for the erector to install them. In most instances, precasters will make their own erection drawings to identify the information for precast concrete units and translate these details into production and erection requirements. A major reason for leaving most panel design and hardware details to the precaster would be that they normally have extensive experience in this field and can choose details suitable for their plant's production and erection techniques.

If reinforcement and connections are not detailed by the architect/engineer, the performance requirements for the design must be clearly identified. The amount of space allowed for connections should also be indicated. It is generally recommended that the architect/engineer define dead, live, wind and seismic loads and describe a section adequate to receive loads, and design the connection for each typical unit. This approach provides a design compatible with structural capability, concrete cover, hardware protection, appearance, and clearance for mechanical services. In addition, it will establish parameters for reasonable modifications which may be suggested by the precaster to suit their production and erection techniques, and at the same time satisfy project design requirements. Design deviations requested by the precaster or erector will be permitted only after the architect/engineer's approval of the proposed change.

Constructor Responsibilities

The general conditions of the construction contract usually state the responsibility of the constructor (usually the general contractor) in coordinating the construction work. The constructor is typically responsible for project schedule, dimensions and quantities, coordination with all other construction trades, and for the adequacy of construction means, methods, techniques, sequences, and procedures of construction, in addition to safety precautions and programs in connection with the project. The precast concrete manufacturer should not proceed with fabrication of any products prior to receiving approval of erection drawings by the constructor.

The constructor should:

1. Be responsible for coordinating all information necessary to produce the precast concrete erection drawings.

2. Review and approve or obtain approval for all precast concrete shop drawings which include the scheme of handling, transporting, and erecting the units, as well as the plan of temporary bracing of the structure. (Handling and transporting responsibilities depend on whether precast concrete is sold F.O.B. plant, F.O.B. jobsite or erected.)

3. Be responsible for the coordination of dimensional interfacing of the precast concrete units with other materials and construction trades.

4. Assure that proper tolerances are maintained in the supporting structure to provide for accurate fit and overall conformity with precast concrete erection drawings.

5. Incorporate on or in the structure contractor's hardware according to a layout or anchor plan supplied by the precaster.

It is the responsibility of the constructor to establish and maintain, at convenient locations, control points and benchmarks in an undisturbed condition for use by the erector until final completion and acceptance of a project.

The constructor must be responsible for coordinating precast concrete erection drawings with shop drawings from other trades so that related items can be transmitted to the designer in one package. The constructor must immediately notify the precaster of any deviations found in dimensions due to plan or construction errors or changes to the structure.

After erection of the precast concrete panels, the constructor should notify the architect for the pre-final inspection of the work. Representatives of the precaster and the erector should be prepared to participate in this inspection tour and answer any questions posed by the architect.

Precast Concrete Manufacturer Responsibillities

The extent of design responsibility vested with the precast concrete manufacturer should be clearly defined by the architect/engineer in the contract documents. Table B1 outlines the typical options.

Most work is covered in Option II Table B1, due to the specialized nature of precast concrete design. Additional design responsibilities for the manufacturer may occur when the architect/engineer uses methods of communication described in Options II and III in Table B1. Option II (b) may occur when no engineer is involved with the architect for precast product design and where doubtful areas of responsibility may exist. Under this option, the precaster should employ or retain a structural engineer experienced in the design of precast concrete, who will ensure the adequacy of those structural aspects of the erection drawings, manufacture and installation for which the manufacturer is responsible. Option III is not yet a common practice but might be used for design of systems buildings or with performance specifications.

Table B1 Design Responsibilities

Contract Information Supplied by Architect/Engineer	Responsibility of the Manufacturer of Precast Concrete Units
OPTION I	
Provide complete drawings and specifications detailing all aesthetic, functional and structural requirements plus dimensions	The manufacturer shall make shop drawings (erection and production drawings), as required, with details as shown by the designer. Modifications may be suggested that, in manufacturer's estimation, would improve the economics, structural soundness or performance of the precast concrete installation. The manufacturer shall obtain specific approval for such modifications. Full responsibility for the product design, including such modifications, shall remain with the architect/engineer. Alternative proposals from a manufacturer should match the required quality and remain within the parameters established for the project. It is particularly advisable for the architect/engineer to give favorable consideration to such proposals if the modifications are suggested so as to conform to the manufacturer's normal and proven procedures.
OPTION II	
Detail all aesthetic and functional requirements but specify only the required structural performance of the precast concrete units. Specified performance shall include all limiting combinations of loads together with their points of application. This information shall be supplied in such a way that all details of the unit can be designed without reference to the behavior of other parts of the structure. The division of responsibility for the design shall be clearly stated in the contract.	The manufacturer has two alternatives: (a) Submit erection and shape drawings with all necessary details and design information for the approval and ultimate responsibility of the architect/engineer. (b) Submit erection and shape drawings for general approval and assume responsibility for part of the structural design, i.e., the individual units but not their effect on the building. Firms accepting this practice may either have a professional engineer in their employ stamp (seal) drawings or commission engineering firms to perform the design and stamp the drawings. The choice between alternatives (a) and (b) shall be decided between the architect/engineer and the manufacturer prior to bidding with either approach clearly stated in the specifications for proper allocation of design responsibility. Experience has shown that divided design responsibility can create contractual problems. It is essential that the allocation of design responsibility is understood and clearly expressed in the contract documents. The second alternative is normally adopted where the architect does not engage a design engineer to assist in the design.
OPTION III	
Cover the required structural performance of the precast concrete units as in Option II and cover all or parts of the aesthetic and functional requirements by performance specifications. Define all limiting factors such as minimum and maximum thickness, depths, weights and any other limiting dimensions. Give acceptable limits of any other requirements not detailed.	The manufacturer shall submit drawings with choices assuming responsibilities as in Option II. The manufacturer completes the design in accordance with the specified performance standards and submits, with the bid, drawings and design information including structural calculations. The manufacturer accepts responsibility for complying with the specified performance standards. After acceptance of the bid, the manufacturer submits shop drawings for review by the architect/engineer and for the approval of the constructor.

APPENDIX C

Finish Samples

PRE-BID SAMPLES

Pre-bid samples, as for all samples, shall only be regarded as a standard for performance within the variations of workmanship and materials to be expected.

Due to individual preferences, differences in sources of supply, or different techniques developed in various plants serving the same area, the architect/engineer should not expect to select one sample and obtain exact matching by all precast concrete producers.

Many architects have developed a practice of making sample selection and approval just prior to bid closing. Thus, for a specific project, the approved precasters' names and corresponding sample code numbers, may be published in an addendum or approval list given in writing to the general contractor.

This practice may result in a slight variation in color, aggregate or texture (but not necessarily quality) from different bidders, since the individual precaster, within specification limits, selects the materials and employs the placing and finishing techniques best suited to their plant operation. The architect/engineer, when making prebid approval of samples part of the specifications, should adhere to the following requirements:

1. Sufficient time should be allowed for the bidder to submit samples or information for approval. Time should also be provided to enable such approvals to be conveyed to the precaster in writing so that the precaster can estimate and submit a bid.

2. Any prebid submittal should be treated in confidence, and the individual producer's solutions and/or techniques protected both before and after bidding.

If the characteristics of submitted prebid samples in any way deviate from the specifications, the manufacturer should make this clear to the architect/engineer when submitting the samples and other required information. For proper evaluation and approval of the samples, the manufacturer should state the reasons for the deviations. These reasons might be the precaster's concern over controlling variation in either color or texture within specified limits. In regard to adequacy of specified materials, concerns about satisfying all conditions of the specifications must be based upon practical plant production requirements, and the performance or weathering of the product in its final location.

The architect/engineer may request data as described in Article 1.5.3 in order to evaluate these deviations. If such deviations and samples are approved, the original project specifications and contract drawings should be changed accordingly by the architect/engineer.

Since some samples are developed for specific projects with particular shapes or other characteristics, while others are more general for simple applications, it is the responsibility of the precaster either to make sure that the architect/engineer does not retain these specific project samples, or that they are clearly marked with respect to limits of application to prevent their use for unsuitable applications.

Some examples of use of samples for specific applications are sandblasted lightweight aggregate units for interior applications, or units that use mixes with lower compressive strength and/or higher absorption percentages for dry, non-corrosive atmospheric conditions, or the use of concrete elements in temporary buildings, such as exhibit halls.

IDENTIFICATION

A file of sample code numbers with all related data should be maintained by the precaster to ensure future duplication of any sample submitted.

MOCKUPS AND PRODUCTION APPROVAL

Aesthetic mockups can offer the opportunity to evaluate the following factors:

1. Range of acceptable appearance in regard to color, texture, details on the exposed face, and uniformity of returns.

2. Sequence of erection.

3. Available methods of bracing units prior to final structural connections being made.

4. Desirability of the method of connection in light of handling equipment and erection procedures.

5. Colors and finishes of adjacent materials (window frames, glass, sealants, etc.)

6. Dimensional accuracy of the precast concrete work and the constructibility of the specified tolerances.

7. The acceptability of the precast concrete panel inside surface finish (where exposed).

8. Available methods for the repair of chips, spalls or other surface blemishes. The mockup will also establish the extent and acceptability of defects and repair work.

9. Suitability of the selected sealers.

10. The weathering patterns or rain run-off on a typical section of precast concrete panel facade.

Mockups should be produced using standard production equipment and techniques. Some important variables that should be controlled as close to actual cast conditions include: retarder coverage rate and method of application, mix design and slump, admixtures, heat of plastic and cured concrete, age, vibration, piece thickness and method of cleaning. This is especially important with light etches which are particularly affected by changing conditions. Special details such as reveal patterns and intersections, corner joinery, drip sections, patterns, color and texture, and other visual panel characteristics should be demonstrated in large production samples for approval. Changes in aggregate orientation, color tone, and texture can easily be noted on full scale mockup panels.

The objective of the mockup sample can also be to demonstrate the more detailed conditions that may be encountered in the project (recesses, reveals, outside/inside corners, multiple finishes, textures, veneers, etc.). This sample may not be fully representative of the exact finishes that can be reasonably achieved during mass production.

Mockup panels should contain typical cast-in-inserts, reinforcement, and plates as required for the project. Handling the mockup panels serves as a check that the stripping methods and lifting hardware will be suitable.

The architect should visit the precast concrete plant for examination and approval (sign and date) of the first production units. To avoid possible later controversies, this approval should precede a release for production. The architect should realize, however, that delays in visiting plants for such approvals may upset normal plant operations and the job schedule. It should be clearly stated in the contract documents how long the production units or the mockup structure should be kept in the plant or jobsite for comparison purposes. It is recommended that the contract documents permit the approved full-sized units to be used in the job installation in the late stages of construction. The units should remain identifiable even on the structure, until final acceptance of the project. The panels should be erected adjacent to each other on the building to allow continued comparison, if necessary.

The face of each sample should contain at least two areas of approved size and shape which have been chipped out and then patched and repaired. The color, texture and appearance of patched areas should match that of adjacent surface; see Article 2.8.

Plant inspection by the architect during panel production is encouraged. This helps assure both the architect and the precaster that the desired end results can and are being obtained.

APPENDIX D

Chuck Use and Maintenance Procedure

Strand chucks are precision pieces of equipment designed to hold thousands of pounds of force. It is critical that inspections and preventive maintenance be performed periodically to insure their performance. The following procedure lists both general rules for using chucks on a bed and for their maintenance in the shop.

A. *Bed Crews*

1. Chucks should always be kept clean. Chucks not on the strand should never be thrown on the ground. All used chucks should be returned to the chuck cleaning bench.

2. To secure strand, always use a chuck that is marked the same size as the strand.

3. Visually inspect each chuck barrel before placing it on the strand. This can be done by glancing down the center of the chuck to be certain all 3 wedges are in alignment and that the chuck is clean.

4. To release a chuck from a piece of strand, simply grasp the chuck and push the barrel forward to remove the wedges. A second alternative would be to force the narrow end of a chuck removal tool into the barrel end of the assembled chuck. This will force the wedges back into the chuck and release it from the strand.

5. If the barrel still does not release, use a grinder to cut the strand approximately two feet on either side of the chuck and take the chuck with the strand in it to the chuck cleaning room so it can be removed properly. The chuck barrel should never be hit with a hammer, rock, rebar or anything else.

7. If the chuck comes off the strand easily, put the wedges back into the barrel.

8. All chucks and wedges should be cleaned, inspected and lubricated before using.

B. *Chuck Maintenance and Inspection Personnel*

1. To remove a chuck which has been brought to the shop on a length of strand, use a chuck removal tool, a vice and a sliding hammer.

   a. Place the chuck removal tool over the strand and push the narrow end up into the barrel end of the chuck.

   b. Place the strand (with the chuck removal tool in place) upright in the vise so that the chuck removal tool rests flat against the vise. Leave the jaws of the vice loose enough so the strand can move between them.

   c. Use the sliding hammer to strike the open end of the chuck. If necessary, another hammer can be used to strike the sliding hammer. Do not strike the chuck barrel itself with any hard object.

   d. Striking the open end with the sliding hammer will force the chuck aganist the removal tool pushing the wedges back into the barrel and jarring the chuck loose. The chuck can then be removed either by pressing the removal tool into the barrel and sliding the chuck off the strand or by disassembling the chuck.

2. To inspect and lubricate the chucks, use the following procedure:

   a. Separate all chucks by size. Inspect one size of chuck completely before placing another size chuck on the bench.

   b. Disassemble all chucks by removing the wedges and the rubber retaining ring.

   c. Put on safety goggles before using the brushes to clean the chucks. Rubber gloves or gloves that will not get caught in the brushes should also be used.

   d. The chuck body can be cleaned with a tapered wire brush.

   e. Before removing the rubber retaining ring, the wedges can be cleaned with a motorized tapered nylon brush. Care must be taken to assure the wire body brush is never used to clean the wedges as damage could result.

   f. After cleaning, remove the rubber retaining ring and inspect the wedges for signs of scratches, chipped threads, and score marks. Examine the rubber retaining rings for signs of splitting or damage. Inspect the barrels for signs of excessive rust, corrosion or dents. *Discard all damaged parts* to assure they are not used in reassembly.

   g. Clean and lubricate the barrels and allow them to dry approximately five minutes. Lubricate the exterior of the jaw assembly.

   h. Reassemble the chucks and store them in a rack designed to keep them clean and out of the weather.

   i. Repeat the process for the next size of chuck.

APPENDIX E

Sample Record Forms

Numerous items which require recordkeeping for confirmation and evaluation are outlined in this quality control manual. The following record forms are suggested for consideration in a quality control program. These are not the only forms needed for operations, but will provide a beginning point for form development. It should be recognized that these are only **SAMPLE** record forms and that a plant may design its own forms to best serve its operation. The importance of any recording form is the information that it contains and not its format. The reports serve as a record of the manufacturing process in case this information is required at some future date.

Items that can be measured quantitatively are to be recorded in numerical terms. Items that must be evaluated subjectively should be rated in a consistent fashion. Items such as length and width measurements can be given a check mark or "O.K." if they are within the tolerances listed in Article 7.1.2.

Where extra attention is required to improve material or product quality, or improve worker quality performance, or to identify matters beyond the control of the worker, remarks or sketches should be used. Remarks can be made on the back of the forms when there is not room on the front side.

The following forms are included:

1. Aggregate Gradation, Material Finer than 200 Sieve, Aggregate Moisture Content, Organic Impurities
2. Batch Plant Scale Check
3. Concrete Tests Report
4. Tensioning Report
5. Inspection Report

QUALITY CONTROL

**AGGREGATE GRADATION
MATERIAL FINER THAN 200
MOISTURE TEST
ORGANIC IMPURITIES**

Date: _____

Inspector: _____

MATERIALS CONTROLS

Sheet No.: _____

AGGREGATES

Applicable Spec.: _____ FINE AGGREGATE   Date Del.: _____

Sieve Size	Weight Retained (g)	% Retained	Cum. % Passing	Specifications % ASTM C33		Remarks Bin ____ Tons Rep. ____	
3/8"							
No. 4							
No. 8							
No. 16							
No. 30							
No. 50							
No. 100							
Pan							
			F.M.				

Applicable Spec.: _____ COARSE AGGREGATE   Date Del.: _____

Sieve Size	Weight Retained (g)	% Retained	Cum. % Passing	Specifications % ASTM C33		Remarks Bin ____ Tons Rep. ____	
1"							
3/4"							
1/2"							
3/8"							
No. 4							
No. 8							
No. 16							

MATERIAL FINER THAN 200 SIEVE (ASTM C117)

_____ Coarse Agg.   _____ Fine Agg.

Supplier _____

Date Del. _____

Original Wt. of Sample _____ (at least 2.5 kg)

Dry Wt. of Orig. Sample _____ = B

Dry Wt. Sample After Washing _____ = C

% Material Finer than 200 Sieve = $\frac{B-C}{C} \times 100 =$ _____ = A

AGGREGATE MOISTURE CONTENT (C566)

Wt. Sample & Container (Wet) _____ (D)

Wt. Sample & Container (Dry) _____ (E)

Wt. of Container _____ (F)

Wt. of Moisture (D-E) _____ (G)

Net Dry Wt. of Sample (E-F) _____ (H)

% Moisture = $\frac{G}{H} \times 100 =$ _____

Speedy Moisture Test _____ %

ORGANIC IMPURITIES IN SAND (C40)

Circle Color of Sodium Hydroxide Solution

Organic Plate No.
1
2
3 (standard)
4
5

Sand Supplier _____

Date _____

QUALITY CONTROL

BATCH PLANT SCALE CHECK

Date: _____

EQUIPMENT CALIBRATION

Sheet No.: _____

AGGREGATE SCALES			CEMENT SCALES			Remarks
Bar	Test Load	Scale Reading	Bar	Test Load	Scale Reading	

☐ Balance Point at Zero at Start of Check

☐ Balance Point Off _____ Pts.   ☐ Under   ☐ Over

☐ Adjustment Required on Bar to Balance

Scale Report to be Completed
On first Day of Each Week

_____
Checked by

MNL-117 3rd Edition

QUALITY CONTROL

CONCRETE TEST REPORT

MATERIALS CONTROLS

Sheet No.: _____

ALL PRODUCTS

Job Name _____ Job No. _____
Date _____ Inspector _____

Mark No.	Pour No.	Specimen No.	Slump (in.)	Air (%)	Unit Wt. (pcf)	Conc. Temp. °F	Amb. Temp. °F	Temp. Time	Date At Break	Time At Break	Curing Duration (hrs.)	Total load	Cylinder or Cube Strength	Remarks

FACE MIX   Mix Designation (sample number) _____   Design Strength _____

BACKUP MIX   Mix Designation (sample number) _____

Impact Hammer (ASTM C805)

Average Rebound Reading (R)   Compressive Strength (psi)

_____   _____
_____   _____
_____   _____
_____   _____

COMMENTS: _____
_____
_____
_____
_____

Concrete Yield Computation (ASTM C138)

Cement	_____	Wt. Container & Concrete _____
Fine Aggregate	_____	Wt. Container _____
Coarse Aggregate	_____	Wt. Concrete _____
Admixtures	_____	Slump _____
Admixture (Pigment)	_____	Entrained Air _____
Water No. Gals. ____ x 8.33 ____		Container Size ____ cu. ft.
Total Wt. per Cu. Yd. (A) ____		Wt. per Cu. Ft. (B) ____

$$\text{Yield} = \frac{A}{B} = _____$$

Air Measurement (ASTM C231)

Aggregate Correction Factor (G)

$h_1 = $ _____

$h_2 = $ _____

$G = h_1 - h_2 = $ _____

Apparent Concrete Air Content (A_1)

$h_1 = $ _____

$h_2 = $ _____

$A_1 = h_1 - h_2 = $ _____

A (Air Content) $= A_1 - G = $ _____

QUALITY CONTROL

TENSIONING REPORT

Date: _____

Inspector: _____

FABRICATION CONTROL

Sheet No.: _____

ALL PRODUCTS

Job Name _____

Job No. _____ Pour No. _____

Bed No. _____ Product _____

Member Identifications _____
_____

Jack Identification _____

PLOT STRANDS ON GRID

(Grid numbered 1–23 vertically, columns A B C D E F G H I J K L M N O P Q R S T U V W X)

Strand Size _____ No. _____ Pattern _____ Strand Brand _____

Type of Tensioning _____ At _____ End

Strand Temperature _____ °F _____ Theoretical Elongation

Theoretical Gauge Reading _____ lbs.

Temperature Correction for Elongation _____ " = * _____ " Total Elongation

Temperature Correction for Gauge Reading _____ lbs. = ** _____ Corrected Reading

LOAD CELLS

Location	Strand Number	Load by Gauge Pressure	Load by Load Cell	Prior to Pour

Strand Pattern Ident.ification	COMPUTED TENSIONING DATA				ACTUAL-LIVE END		NOTES
	Total Elongation	Pretension	Net Elongation	Gauge Pressure	Net Elongation	Gauge Pressure	

QUALITY CONTROL

INSPECTION REPORT
BUILDING PRODUCTS
FORM SET-UP, CASTING AND FINISHED PRODUCT

Date Produced: _____

_____ Inspector

Date of Post Inspection: _____

Mold No.: _____
Job No.: _____
Product: _____
Job Name: _____
Release Agent: _____

ID.	Job No.					
	Member Mark No.					
SET-UP DETAILS	Form Conditions & Cleanliness					
	Seams Sealed					
	Retarders					
	Design Length (Height)					
	Form Set-Up Length (Height)					
	As-Cast Length (Height)					
	Design Width					
	Form Set-Up Width					
	As-Cast Width					
	Design Depth (Thickness)					
	Form Set-Up Depth (Thickness)					
	As-Cast Depth (Thickness)					
	Out of Square					
	Block Outs					
	Squareness of Openings					
	Veneer Alignment					
	End & Edge Details					
	Reinforcement					
	Insulation					
	Reglets					
	Rustications					
	Haunches (Corbels)					
	Plates & Inserts					
	Lifting Devices					
CASTING	Vibration Rating					
	Workability					
	Top Finish (Wet)					
FINISHED PRODUCT	Top Finish (Dry)					
	Bottom Finish (Dry)					
	Surface Textures					
	Color Uniformity					
	Cracks or Spalls					
	Out of Square (Max.)					
	Camber or Deflection					
	Warpage					
	Bowing					
	Exposed Reinforcing or Chairs					
	Plates & Inserts					
	Chamfers & Radius Quality					
	Openings & Block Outs					
	Lifting Devices					
	Panel Sealer Applied					
YARDAGE	Blocking					
	Finishing					
	Patching & Cleaning					
	Date Approval Stamp Applied					
LOADING	Blocking					
	Field Patching Required					
	Tie Downs					
	Driver Instructions					

☐ Product Accepted   ☐ Product Rejected   Reason _____

USE BACKSIDE OF SHEET FOR REMARKS AND SKETCHES

APPENDIX F

PCI Plant Certification Program

In 1958, the Precast/Prestressed Concrete Institute recognized that a dedicated effort to produce quality products was necessary and that such an effort would play an integral role in the development of the precast and prestressed concrete industry. Following the publishing of several quality and standards manuals, in 1967, the Institute inaugurated its Plant Certification Program which has continued to grow since that time. The program is now mandatory for all PCI members and the expense is borne primarily by the individual participating plants. Unannounced audits are performed twice a year by a nationally recognized structural engineering consultant to assure impartial and uniform evaluations. Auditors are graduate and professional structural engineers with experience in design, construction, and manufacture of precast and prestressed concrete products.

The Plant Certification Program was expanded in 1970 to include production of architectural precast concrete and again in 1979 to include glass fiber reinforced concrete products.

In 1988, PCI implemented a further expansion of the Plant Certification Program to require certification by the type of products manufactured. This refinement of the program requires a more product-specific inspection and evaluation of a plant's specialized capabilities. Plants may be certified in up to four product groups:

Group A	Group B	Group C	Group G
Architectural Concrete	Bridges	Commercial (Structural)	Glass Fiber Reinforced Concrete

Group A includes two (2) categories; category AT for miscellaneous Architectural Trim elements and category A1 for Architectural Precast Concrete.

Within product groups B and C are four (4) categories that identify product types that more effectively convey the expertise and capability of the individual plant. The categories reflect differences in the ways in which the products are produced.

PCI's Plant Certification Program is recognized as a quality assurance agency by the following:

- International Conference of Building Officials (ICBO)
- Building Officials and Code Administrators (COBA)
- Southern Building Code Congress International, Inc. (SBCC)

PCI Plant Certification is listed in the following specifications:

- Master Specification — American Institute of Architects (AIA)
- Naval Facilities Engineering Command
- U.S. Army Corps of Engineers

PCI Plant Certification is highly endorsed by the Federal Highway Administration and is now required by more than twelve (12) individual state departments of transportation.

Statement of Purpose

The certification of a producing plant by PCI indicates the plant has a system of quality in place starting with management. Further, the quality system extends to all areas of plant operations including a thorough and well documented quality control program. Certification indicates that plant practices are in conformance with time-tested industry standards. It means that the plant regularly demonstrates the capability to produce quality products of the type in which they are certified.

PCI certification does not guarantee the quality of specific precast concrete products.

Qualification of Certification Applicants

Applicants for PCI Plant Certification must be producers of precast and/or prestressed concrete products, and must have been in production for twelve (12) months prior to the Initial audit. Participants may be either PCI producer members or non-members.

Operation of Certification Program

All plants are audited at least twice each year. Audits are unannounced and are usually of two days duration. New participants are audited three times during the first year.

Auditors compile numerical grades and prepare an in-depth written audit report.

Grading is organized in keeping with the Divisions in this manual, MNL-117, and a specific numerical grade is required for certification. The audit covers all aspects of production and quality control as well as engineering and general plant practices. During each audit, stored products are reviewed and selected panels are measured and compared to shop drawings to verify conformance with tolerances. The product evaluations performed by in-house quality control personnel are also reviewed to determine if regular monitoring is correct and accurate.

Audits and grading of structural precast/prestressed concrete are based on MNL-116. Audits and grading of glass fiber reinforced concrete are based on MNL-130.

A list of Certified Plants is published quarterly by PCI. Failure to meet standards results in loss of certification. In this event, certification may be regained only by passing a new, special audit. Grades cannot be negotiated or mitigated. Specific grades are confidential and are not released.

Plant Certification continues to be rated by users as a highly effective tool for auditing their internal quality assurance programs. For further information about the program and a list of Certified Plants, please contact:

Director of Certification Programs
Precast/Prestressed Concrete Institute
175 West Jackson Boulevard
Chicago, Illinois 60604
(312) 786-0300

GUIDE QUALIFICATION SPECIFICATION ARCHITECTURAL PRECAST

The architectural precast concrete manufacturing plant shall be certified by the Precast/Prestressed Concrete Institute PCI Plant Certification Program. Manufacturer shall be certified at time of bidding. Certification shall be in product group and category [select one or both: AT-Miscellaneous Architectural Trim Units; A1-Architectural Precast Concrete].

APPENDIX G

Reference Literature

This manual and its commentary refer to many standards and outline recommendations based on the available body of knowledge involving precast and prestressed concrete. This appendix provides a basic outline of applicable standards and reference material. It is essential that a production facility's personnel be furnished with current reference literature and be encourage to read and utilize it.

A minimum reference list should include applicable and current publications of the American Society for Testing and Materials; the American Concrete Institute; the Precast/Prestressed Concrete Institute; the Portland Cement Association; and similar agencies having pertinent applicable specifications dealing with manufacture of precast concrete.

American Society for Testing and Materials (ASTM)
100 Barr Harbor Drive
West Conshohocken, Pennsylvania 19428-2959

The ASTM Book of Standards contains specifications and test methods for most of the materials and standard practices used in the production of architectural precast concrete. They also contain specifications and methods of test for related materials.

The following is a list of individual standards which can be ordered in section groups published by the society.

ASTM Designation	Title
A 27/A27M	Specification for Steel Castings, Carbon, for General Application
A 36/A36M	Specification for Carbon Structural Steel
A 47	Specification for Ferritic Malleable Iron Castings
A 47M	Specification for Ferritic Malleable Iron Castings (Metric)
A 82	Specification for Steel Wire, Plain for Concrete Reinforcement
A 108	Specification for Steel Bars, Carbon, Cold Finished, Standard Quality
A 123	Specification for Zinc (Hot-Dip Galvanized) Coatings on Iron and Steel Products
A 143	Practice for Safeguarding Against Embrittlement of Hot-Dip Galvanized Structural Steel Products and Procedure for Detecting Embrittlement
A 153	Specification for Zinc Coating (Hot-Dip) on Iron and Steel Hardware
A 184/A184M	Specification for Fabricated Deformed Steel Bar Mats for Concrete Reinforcement
A 185	Specification for Steel Welded Wire Fabric, Plain, for Concrete Reinforcement
A276	Specification for Stainless and Heat-Resisting Steel Bars and Shapes
A 283/A283M	Specification for Low and Intermediate Tensile Strength Carbon Steel Plates
A 307	Specification for Carbon Steel Bolts and Studs, 60,000 PSI Tensile Strength
A325	Specification for Structural Bolts, Steel, Heat Treated, 120/105 ksi Minimum Tensile Strength
A 370	Test Methods and Definitions for Mechanical Testing of Steel Products
A 416/A416M	Specification for Steel Strand, Uncoated Seven-Wire for Prestressed Concrete
A 421	Specification for Uncoated Stress-Relieved Steel Wire for Prestressed Concrete
A 496	Specification for Steel Wire, Deformed, for Concrete Reinforcement
A 497	Specification for Steel Welded Wire Fabric, Deformed, for Concrete Reinforcement
A 500	Specification for Cold-Formed Welded and Seamless Carbon Steel Structural Tubing in Rounds and Shapes
A 572/A572M	Specification for High-Strength Low-

	Alloy Columbium-Vanadium Structural Steel
A 615/A615M	Specification for Deformed and Plain Billet-Steel Bars for Concrete Reinforcement
A 616/A616M	Specification for Rail-Steel Deformed and Plain Bars for Concrete Reinforcement
A 617/A617M	Specification for Axle-Steel Deformed and Plain Bars for Concrete Reinforcement
A 641	Specification for Zinc-Coated (Galvanized) Carbon Steel Wire
A 641M	Specification for Zinc-Coated (Galvanized) Carbon Steel Wire (Metric)
A 666	Specification for Austenitic Stainless Steel, Sheet, Strip, Plate and Flat Bar
A 675/A675M	Specification for Steel Bars, Carbon, Hot-Wrought, Special Quality, Mechanical Properties
A 706/A706M	Specification for Low-Alloy Steel Deformed Bars for Concrete Reinforcement
A 722	Specification for Uncoated High-Strength Steel Bar for Prestressing Concrete
A 767/A767M	Specification for Zinc-Coated (Galvanized) Steel Bars for Concrete Reinforcement
A 775/A775M	Specification for Epoxy-Coated Reinforcing Steel Bars
A 780	Practice for Repair of Damaged and Uncoated Areas of Hot-Dip Galvanized Coatings
A 884	Specification for Epoxy-Coated Steel Wire and Welded Wire Fabric for Reinforcement
A 934/A934M	Specification for Epoxy-Coated Prefabricated Steel Reinforcing Bars
B 633	Specification for Electrodeposited Coatings of Zinc on Iron and Steel
B 766	Specification for Electrodeposited Coatings of Cadmium
C 29	Test Method for Unit Weight and Voids in Aggregate
C 31	Practice for Making and Curing Concrete Test Specimens in the Field
C 33	Specification for Concrete Aggregates
C 39	Test Method for Compressive Strength of Cylindrical Concrete Specimens
C 40	Test Method for Organic Impurities in Fine Aggregates for Concrete
C 42	Test Method for Obtaining and Testing Drilled Cores and Sawed Beams of Concrete
C 67	Test Methods of Sampling and Testing Brick and Structural Clay Tile
C 70	Test Method for Surface Moisture in Fine Aggregate
C 88	Test Method for Soundness of Aggregates by Use of Sodium Sulfate or Magnesium Sulfate
C 94	Specification for Ready-Mixed Concrete
C 109	Test Method for Compressive Strength of Hydraulic Cement Mortars (Using 2-in. or 50-mm Cube Specimens)
C 117	Test Method for Materials Finer than No. 200 (75-μm) Sieve in Mineral Aggregates by Washing
C 125	Terminology Relating to Concrete and Concrete Aggregates
C 127	Test Method for Specific Gravity and Absorption of Coarse Aggregate
C 128	Test Method for Specific Gravity and Absorption of Fine Aggregate
C 131	Test Method for Resistance to Degradation of Small-Size Coarse Aggregate by Abrasion and Impact in the Los Angeles Machine
C 136	Test Method for Sieve Analysis of Fine and Coarse Aggregates
C 138	Test Method for Unit Weight, Yield, and Air Content (Gravimetric) of Concrete
C 142	Test Method for Clay Lumps and Friable Particles in Aggregates
C 143	Test Method for Slump of Hydraulic Cement Concrete
C 144	Specification for Aggregate for

	Masonry Mortar	C 457	Test Method for Microscopical Determination of Parameters of the Air-Void System in Hardened Concrete
C 150	Specification for Portland Cement		
C 171	Specification for Sheet Materials for Curing Concrete	C 469	Test Method for Static Modulus of Elasticity and Poisson's Ratio of Concrete in Compression
C 172	Practice for Sampling Freshly Mixed Concrete	C 470	Specification for Molds for Forming Concrete Test Cylinders Vertically
C 173	Test Method for Air Content of Freshly Mixed Concrete by the Volumetric Method	C 494	Specification for Chemical Admixtures for Concrete
C 185	Test Method for Air Content of Hydraulic Cement Mortar	C 566	Test Method for Total Moisture Content of Aggregate by Drying
C 188	Test Method for Density of Hydraulic Cement	C 567	Test Method for Unit Weight of Structural Lightweight Concrete
C 191	Test Method for Time of Setting of Hydraulic Cement by Vicat Needle	C 578	Specification for Rigid, Cellular Polystyrene Thermal Insulation
C 192	Practice for Making and Curing Concrete Test Specimens in the Laboratory	C 586	Test Method for Potential Alkali Reactivity of Carbonate Rocks for Concrete Aggregates (Rock Cylinder Method)
C 204	Test Method for Fineness of Hydraulic Cement by Air Permeability Apparatus		
C 227	Test Method for Potential Alkali Reactivity of Cement-Aggregate Combinations (Mortar-Bar Method)	C 591	Specification for Unfaced Preformed Rigid Cellular Polyisocyanurate Thermal Insulation
		C 595	Specification for Blended Hydraulic Cements
C 231	Test Method for Air Content of Freshly Mixed Concrete by the Pressure Method	C 617	Practice for Capping Cylindrical Concrete Specimens
C 233	Test Method for Air-Entraining Admixtures for Concrete	C 618	Specification for Coal Fly Ash and Raw or Calcined Natural Pozzolan for Use as a Mineral Admixture in Concrete
C 260	Specification for Air-Entraining Admixtures for Concrete		
C 295	Guide for Petrographic Examination of Aggregates for Concrete	C 641	Test Method for Staining Materials in Lightweight Concrete Aggregates
C 309	Specification for Liquid Membrane-Forming Compounds for Curing Concrete	C 642	Test Method for Specific Gravity, Absorption, and Voids in Hardened Concrete
C 330	Specification for Lightweight Aggregates for Structural Concrete	C 666	Test Method for Resistance of Concrete to Rapid Freezing and Thawing
C 342	Test Method for Potential Volume Change of Cement-Aggregate Combinations	C 685	Specification for Concrete Made by Volumetric Batching and Continuous Mixing
C 370	Test Method for Moisture Expansion of Fired Whiteware Products	C 702	Practice for Reducing Samples of Aggregate to Testing Size
C 403	Test Method for Time of Setting of Concrete Mixtures by Penetration Resistance	C 805	Test Method for Rebound Number of Hardened Concrete
		C 845	Specification for Expansive Hydraulic Cement

C 881	Specification for Epoxy-Resin-Base Bonding Systems for Concrete
C 917	Test Method for Evaluation of Cement Strength Uniformity from a Single Source
C 979	Specification for Pigments for Integrally Colored Concrete
C 989	Specification for Ground Granulated Blast-Furnace Slag for Use in Concrete and Mortars
C 1017	Specification for Chemical Admixtures for Use in Producing Flowing Concrete
C 1059	Specification for Latex Agents for Bonding Fresh to Hardened Concrete
C 1064	Test Method for Temperature of Freshly Mixed Portland Cement Concrete
C 1077	Practice for Laboratories Testing Concrete and Concrete Aggregates for Use in Construction and Criteria for Laboratory Evaluation
C 1105	Test Method for Length Change of Concrete Due to Alkali-Carbonate Rock Reaction
C 1126	Specification for Faced or Unfaced Rigid Cellular Phenolic Thermal Insulation
C 1157/C1157M	Performance Specification for Blended Hydraulic Cement
C 1218	Test Method for Water-Soluble Chloride in Mortar and Concrete
C 1231	Practice for Use of Unbonded Caps in Determination of Compressive Strength of Hardened Concrete Cylinders
C 1240	Specification for Silica Fume for Use in Hydraulic-Cement Concrete and Mortar
C 1252	Test Methods for Uncompacted Void Content of Fine Aggregate (as Influenced by Particle Shape, Surface Texture and Grading)
C 1260	Test Method for Potential Alkali Reactivity of Aggregates (Mortar-Bar Method)
D 75	Practice for Sampling Aggregates
D 4791	Test Method for Flat or Elongated Particles in Coarse Aggregate
E 4	Practices for Force Verification of Testing Machines
E 11	Specification for Wire Cloth Sieves for Testing Purposes
E 105	Practice for Probability Sampling of Materials
E 329	Specification for Agencies Engaged in the Testing and/or Inspection of Materials Used in Construction

For all materials and equipment used in the manufacture of precast and prestressed concrete, for which an appropriate ASTM designation has not been developed, manufacturer's specifications and directions should be available. Such materials and equipment should be used only when they have been shown by tests to be adequate for the purpose intended and their usage has been approved by the purchasing entity.

American Concrete Institute
P.O. Box 9094
Farmington Hills, MI 48333

1. Manual of Concrete Inspection, SP-2
2. Manual for Concrete Practice

    Part 1 Materials and General Properties of Concrete

    Part 2 Construction Practices and Inspection of Pavements

    Part 3 Use of Concrete in Building - Design, Specifications, and Related Topics

    Part 4 Bridges, Substructures, Sanitary, and Other Special Structures; Structural Properties

    Part 5 Masonry; Precast Concrete; Special Processes

These volumes contain accepted ACI Standards including the Building Code requirements and appropriate publications covering all aspects of concrete proportioning, batching, mixing, placing, and curing. They should be available in all precast plants. Some of the more pertinent recommended practices and guides are as follows:

ACI Designation	Title
116R	Cement and Concrete Terminology
117R	Standard Specifications for Tolerances for Concrete Construction and Materials

201.2R	Guide to Durable Concrete
211.1	Standard Practice for Selecting Proportions for Normal, Heavyweight, and Mass Concrete
211.2	Standard Practice for Selecting Proportions for Structural Lightweight Concrete
211.3	Standard Practice for Selecting Proportions for No-Slump Concrete
212.3R	Chemical Admixtures for Concrete
212.4	Guide for the Use of High-Range Water Reducing Admixtures (Superplasticizers) in Concrete
213R	Guide for Structural Lightweight Aggregate Concrete
214	Recommended Practice for Evaluation of Strength Test Results of Concrete
214.3R	Simplified Version of the Recommended Practice for Evaluation of Strength Test Results of Concrete
221.R	Guide for Use of Normal Weight Aggregates in Concrete
224.1R	Causes, Evaluation, and Repair of Cracks in Concrete Structures
225R	Guide to the Selection and Use of Hydraulic Cements
301	Specifications for Structural Concrete for Buildings
303R	Guide to Cast-in-Place Architectural Concrete Practice
304R	Guide for Measuring, Mixing, Transporting, and Placing Concrete
304.5	Batching, Mixing, and Job Control of Lightweight Concrete
305R	Hot Weather Concreting
306R	Cold Weather Concreting
308	Standard Practice for Curing Concrete
309R	Guide for Consolidation of Concrete
309.1R	Behavior of Fresh Concrete During Vibration
309.2R	Identification and Control of Consolidation-Related Surface Defects in Formed Concrete
311.5R	Batch Plant Inspection and Field Testing of Ready-Mixed Concrete
318	Building Code Requirements for Structural Concrete and Commentary
363R	State-of-the-Art Report on High-Strength Concrete
423.3R	Recommendations for Concrete Members Prestressed with Unbonded Tendons
439.4R	Steel Reinforcement — Physical Properties and U.S. Availability
517.2R	Accelerated Curing of Concrete at Atmospheric Pressure

Precast/Prestressed Concrete Institute
175 W. Jackson Blvd.
Chicago, IL 60604

PCI Designation	Title
MNL-116	Manual for Quality Control for Plants and Production of Precast and Prestressed Concrete Products
MNL-119	PCI Drafting Handbook - Precast and Prestressed Concrete
MNL-120	PCI Design Handbook - Precast and Prestressed Concrete
MNL-122	Architectural Precast Concrete
SLP-122	PCI Safety and Loss Prevention Manual
JR-307	Tolerances for Precast and Prestressed Concrete
TN-3	Efflorescence on Precast Concrete

Portland Cement Association
5420 Old Orchard Road
Skokie, IL 60077

PCA Designation	Title
EB 1	Design and Control of Concrete Mixtures
IS 214	Removing Stains and Cleaning Concrete Surfaces

American Welding Society
550 N.W. LeJeune Rd.
P.O. Box 351040
Miami, FL 33135

AWS Designation	Title
A5.1	Specification for Carbon Steel Electrodes for Shielded Metal Arc

	Welding
A5.4	Specification for Stainless Steel Electrodes for Shielded Metal Arc Welding
A5.5	Specification for Low Alloy Steel Covered Arc Welding Electrodes
A5.18	Specification for Carbon Steel Filler Metals for Gas Shielded Arc Welding
A5.20	Specification for Carbon Steel Electrodes for Flux Cored Arc Welding
A5.28	Specification for Low Alloy Steel Filler Metals for Gas Shielded Arc Welding
A5.29	Specification for Low Alloy Steel Electrodes for Flux Cored Arc Welding
B1.11	Guide for Visual Inspection of Welds
B2.1	Standard for Welding Procedure and Performance Qualification
C5.4	Recommended Practices for Stud Welding
D1.1	Structural Welding Code - Structural Steel
D1.4	Structural Welding Code - Reinforcing Steel
QC1	Specification for Qualification and Certification of Welding Inspectors
Z49.1	Safety in Welding, Cutting and Allied Processes

Canadian Standards Association
178 Rexdale Boulevard
Etobicoke (Toronto)
Ontario, Canada M9W 1R3

CSA Designation	Title
A23.1/A23.2	Concrete Materials and Methods of Concrete Construction/Methods of Test for Concrete
A23.2/14A	Potential Expansivity of Aggregates (Procedure for Length Change due to Alkali-Aggregate Reaction in Concrete Prisms)

National Ready Mixed Concrete Association
900 Spring Street
Silver Spring, MD 20910

NRMCA Designation	Title
Pub. #102	Recommended Guide Specifications for Batching Equipment and Control Systems in Concrete Batch Plants

- Concrete Plant Standards of the Concrete Plant Manufacturers Bureau
- Concrete Plant Mixer Standards of Plant Mixer Manufacturers Division, CPMB
- Truck Mixer and Agitator Standards of the Truck Mixer Manufacturers Bureau
- Certification of Ready Mixed Concrete Production Facilities

U.S. Bureau of Reclamation
Denver Federal Center
Denver, CO 80225

- Concrete Manual

Concrete Reinforcing Steel Institute
933 N. Plum Grove Road
Schaumburg, IL 60173

CRSI Designation	Title

- Guidelines for Inspection and Acceptation of Epoxy-Coated Reinforcing Bars at the Job Site
- Field Handling Techniques for Epoxy-Coated Rebar at the Job Site
- Fusion Bonded Epoxy Coating Applicator Plant Certification Program

APPENDIX H

Sample Tensioning Calculations

A. Pretensioned Straight Strand

The following example details the method for calculating the elongation of a straight strand in an abutment anchorage setup. Adjustments for abutment rotation, dead end seating loss, live end seating loss, and temperature variation are shown.

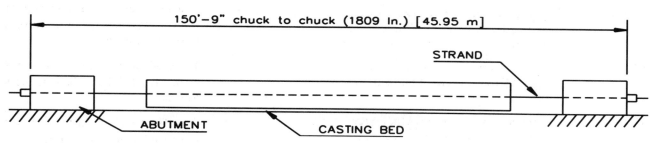

First, the necessary material data and bed setup information must be established.

1. Size and type of strand: 3/8 in. [9.53 mm] diameter, 270K [1860 MPa]
2. Physical characteristics of strand:
   From the mill certificate supplied by the manufacturer,
   $A = 0.0845$ in.^{2} [54.5 mm^2]
   $E = 28,350,000$ psi [195,473. MPa]
   The average values being used by the plant are,
   $A = 0.0850$ in.^{2} [54.8 mm^2]
   $E = 28,500,000$ psi [196,507 MPa]
   Comparing the actual values to the averages,

$$\frac{AE}{A_a E_a} = \frac{0.0845 \times 28,350,000}{0.0850 \times 28,500,000} \times 100 = 98.9\%$$

$$\left[\frac{54.5 \times 195,473}{54.8 \times 196,507} \times 100 = 98.9\% \right]$$

Since AE is within 2.5% of average, the average values can be used in the calculations. *Note that if AE was not within 2.5% of average, this does not indicate that the strand is not in compliance with specification, but only that the actual values from the mill certificate must be used in calculations.*

3. Initial tension of 1,500 lbs [6,672 N] has proven adequate on strand in this bed in the past.

4. Strand is to be stressed to 75% of ultimate,
   23,000 lbs × 0.75 = 17,250 lbs
   [102,304 N × 0.75 = 76,728 N]

Corrections to Tensioning:

a. Abutment Rotation
   Based on ongoing monitoring of abutments under various strand patterns, the abutments are expected to rotate inward under load 1/8 in. [3 mm] each, for a total correction of 1/4 in. [6 mm].

b. Dead End Seating Loss
   Based on ongoing monitoring, seating after initial tension is applied is expected to be 1/8 in. [3 mm].

c. Live End Seating Loss
   Expect 3/8 in. [10 mm] based on past history. Over pull of 3/8 in. [10 mm] is required.

d. Temperature Variation
   Strands will have a temperature of 50 F [10 C] when stressed. The concrete is expected to be at 85 F [29 C] based on current production monitoring, giving an anticipated change of +35 F [19 C].

Self-stressing forms do not require corrections for abutment rotation or temperature variation; however, they do require a correction for elastic shortening of the bed. Elongation should be increased by the average value of the bed shortening, determined by historical data from observations during tensioning, and the strand force increased accordingly.

Tensioning Computations:

$$\text{Basic Elongation} = \frac{(\text{Force required beyond initial tension})(\text{Length of strand between anchorages})}{(\text{Area of strand})(\text{Modulus of elasticity})}$$

$$\text{Basic Elongation} = \frac{(17{,}250 - 1{,}500) \text{ lbs} \times 1809 \text{ in.}}{0.085 \text{ in.}^2 \times 28{,}500{,}000 \text{ psi}} = 11.76 \text{ in.}$$

$$\left[\frac{(76{,}728 - 6{,}672) \text{ N} \times 45.95 \text{ m}}{54.8 \text{ mm}^2 \times 196{,}507 \text{ MPa}} = 0.299 \text{ m} \right]$$

Theoretical Elongation = Basic Elongation combined with appropriate corrections.

Computation of Corrections to Tensioning:

Based on the assumption that elongation will be measured relative to abutment or live end chuck bearing on the abutment, the following will be required.

a. Abutment Rotation: Add 1/4 in. [6 mm] to elongation. No adjustment to force will be made.
  Note that when a large number of strand are used, or when using a multi-strand tensioning scheme, the force lost in the individual strands can be significant, and a force correction would be required.

b. Dead End Seating: Add 1/8 in. [3 mm] to elongation. No adjustment to force is required.

c. Temperature Adjustment (required for variations of 25 F [14 C] or greater)
  Adjust 1% per 10 F [5.5 C] variation. Since the strand will be warmed as the concrete is placed, over-pull is required.

$$\text{Elongation Adjustment} = \frac{17{,}250 \times 1{,}809}{0.085 \times 28{,}500{,}000} \times 0.035 = 0.451 \text{ in.}$$

Force = 17,250 × 1.035 = 17,854 lbs

$$\left[\text{Elongation Adjustment} = \frac{76{,}728 \text{ N} \times 45.95 \text{ m}}{54.8 \text{ mm}^2 \times 196{,}507 \text{ MPa}} \times 0.035 = 0.011 \text{ m} \right]$$

$$[\text{Force} = 76{,}728 \text{ N} \times 1.035 = 79{,}413 \text{ N}]$$

d. Live End Seating : Over pull by 3/8 in. [10 mm]. Adjust force accordingly.

$$\frac{11.76 \text{ in.}}{15{,}750 \text{ lbs}} = \frac{0.375 \text{ in.}}{\text{Added Force}} \quad \text{therefore,} \quad \text{Added Force} = \frac{0.375 \times 15{,}750}{11.76} = 502 \text{ lbs}$$

$$\left[\text{Added Force} = \frac{10 \text{ mm} \times 70{,}056 \text{ N}}{300 \text{ mm}} = 2{,}335 \text{ N} \right]$$

Total Load Required = 17,854 + 502 = 18,356 lbs   [79.413 + 2.335 = 81.748 kN]

Elongation Computation Summary:

	Gross Theoretical Elongation		Net Theoretical Elongation	
Basic Elongation	11.76 in.	299 mm	11.76 in.	299 mm
Abutment Rotation	0.25	6	0.25	6
Dead End Seating Loss	0.125	3	0.125	3
Temperature Adjustment	0.451	11	0.451	11
Live End Seating Loss	0.375	10	0.0	0
Total Elongation	12.961 in.	329 mm	12.586 in.	319 mm
Rounded	13 in.	330 mm	12⅝ in.	320 mm
Tolerance Limits	−5% = 12⅜ in. [312 mm] +5% = 13⅝ in. [346 mm]		−5% = 12 in. [304 mm] +5% = 13¼ in. [336 mm]	

Use Gross Theoretical Elongation for monitoring travel of ram, and compare to 18,356 lb [81.748 kN] force. Use Net Theoretical Elongation for comparison, after seating live end chuck, against movement of mark on strand from initial tension reference.

Note that if the required temperature differential was greater, the total force required during jacking (in this case, 18,356 lbs) would exceed 80% of the ultimate tensile strength (0.80 × 23,000 = 18,400 lbs [81.843 kN]). This would require an allowance for temporary strand stresses to exceed the 80% limit, or other means to control the temperature differential.

B. Post-tensioned Panel Using Straight Single Strand Tendon

The following example details the method for calculating the elongation of a straight, greased and plastic coated (unbonded) strand. Adjustments for anchor wedge seating, elastic shortening, and friction losses are shown.

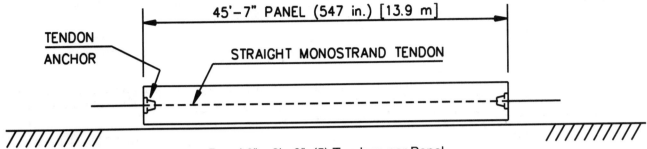

Panel 6" x 8' - 0", (5) Tendons per Panel

First, the necessary material data and tensioning setup information must be established.

1. Size and type of strand: 1/2 in. [12.7 mm] diameter, 270 K [1,860 MPa]

2. Physical characteristics of strand:
   From the mill certificate supplied by the manufacturer,
   $A = 0.1536$ in.^{2} [99.1 mm^2]
   $E = 28,720,000$ psi [198,024 MPa]
   The average values being used by the plant are,
   $A = 0.1530$ in.^{2} [98.71 mm^2]
   $E = 28,500,000$ psi [196,500 MPa]
   Comparing the actual values to the averages,

   $$\frac{AE}{A_a E_a} = \frac{0.1536 \times 28,720,000}{0.1530 \times 28,500,000} \times 100 = 101.2\%$$

   $$\left[\frac{99.1 \text{ mm}^2 \times 198,024 \text{ MPa}}{98.71 \text{ mm}^2 \times 196,500 \text{ MPa}} \times 100 = 101.2\% \right]$$

   Since AE is within 2.5% of average, the average values can be used in the calculations. *Note that if AE was not within 2.5% of average, this does not indicate that the strand is not in compliance with specification, but only that the actual values from the mill certificate must be used in calculations.*

3. From information supplied by the tendon manufacturer,
   κ (wobble friction coefficient) = 0.0014 per foot of tendon [0.0046 per meter]
   μ (curvature friction coefficient) = 0.05

4. Use initial tension of 3,000 lbs. [13.344 kN]

5. Strand is to be stressed to 70% of ultimate,
   41,300 lbs x 0.70 = 28,910 lbs
   [183.7 kN x 0.70 = 128.6 kN]

Corrections to Tensioning:

a. Dead End Wedge Seating Loss
   Based on ongoing monitoring, seating after initial tension is applied is expected to be 1/16 in. [1.5 mm].

b. Live End Wedge Seating Loss
   Expect 3/16 in. [5 mm] based on past history. Over pull of 3/16 in. [5 mm] is required.

c. Elastic Shortening of Panel

   $$ES = 0.5 \, (f_{cpa}) \, \frac{L}{E_c}$$

   in which f_{cpa} is the average compressive stress in the concrete along the member length at the center of gravity of the tendons immediately after tensioning.

   $$f_{cpa} = \frac{P}{A} = \frac{5 \times 28,910}{6 \text{ in.} \times 96 \text{ in.}} = 251 \text{ psi}$$

   $$\left[\frac{5 \times 128.6}{(0.152 \text{ m}) \times (2.44 \text{ m})} = 1.73 \text{ MPa} \right]$$

   $E_c = 33 \, (150)^{1.5} \left(\sqrt{3,000 \text{ psi}} \right) = 3,320,561$ psi

   $0.043 \, (2,400 \text{ kg/m}^3)^{1.5} \left(\sqrt{20.7 \text{ MPa}} \right) = 23,000$ MPa

   $$ES = 0.5 \times 251 \times \frac{547 \text{ in.}}{3,320,561 \text{ psi}} = 0.021 \text{ in.}$$

   $$\left[0.5 \times (1.73 \text{ MPa}) \times \frac{13.9 \text{ m}}{23,000 \text{ MPa}} = 0.5 \text{ mm} \right]$$

   The calculation of elastic shortening is shown here for demonstration only. Generally, with the low levels of prestressing used in these panels, and their relatively short length, elastic shortening can be neglected for tensioning calculations.

d. Friction Losses:

Friction in the tendon system will result in a reduced strand stress at the dead (non-jacking) end of the tendon. Thus, some over pull is required to ensure that the average strand stress equals the design value. When friction losses are high, it is recommended that sequential jacking at both ends of the tendon be used to reduce the possibility of overstressing the strand at the live end.

$$\frac{P_D}{P_S} = e^{-(\kappa L + \mu \alpha)}$$

where P_D equals the force in the strand at the dead end, P_S equals the force in the strand at the live end, L is the tendon length, in feet, between anchorages, and α is the total angular change of the tendon profile, in radians, between the anchorages. For this example,

$\alpha = 0$ for a straight tendon, and $\quad \frac{P_D}{P_S} = e^{-(0.0014 \times 45.583 \text{ ft})} = 0.938 \quad \left[e^{-(0.0046 \times 13.9 \text{ m})} = 0.938 \right]$

Average strand force = (1 + 0.938)/2 = 0.969, loss = 3.1%.

$$\text{Basic Elongation} = \frac{\text{(Force required beyond initial tension) (Length of strand between anchorages)}}{\text{(Area of strand) (Modulus of elasticity)}}$$

$$\text{Basic Elongation} = \frac{(28,910 - 3,000) \text{ lbs} \times 547 \text{ in.}}{0.153 \text{ in.}^2 \times 28,500,000 \text{ psi}} = 3.25 \text{ in.}$$

$$\left[\frac{(128.6 - 13.3) \text{ kN} \times 13.9 \text{ m}}{(98.71 \times 10^{-6} \text{ m}^2 \times 196,500 \text{ MPa} \times 1000)} = 0.0827 \text{ m} \right]$$

Theoretical Elongation = Basic Elongation combined with appropriate corrections.

a. Dead End Seating: Add 1/16 in. [1.5 mm] to elongation. No adjustment to force is required.

b. Live End Seating: Over pull by 3/16 in. [5 mm]. Adjust force accordingly.

$$\frac{3.25 \text{ in.}}{25,910 \text{ lbs}} = \frac{0.1875 \text{ in.}}{\text{Added Force}} \text{, therefore, Added Force} = \frac{0.1875 \times 25,910}{3.25} = 1,495 \text{ lbs}$$

$$\left[\frac{(5 \text{ mm}) \times (115.3 \text{ kN})}{82.7 \text{ mm}} = 6.97 \text{ kN} \right]$$

c. Neglect Elastic Shortening

d. Friction Losses
The total force at the live end must be increased to compensate for friction losses.

Elongation Adjustment = 3.25 × 0.031 = 0.101 in. [82.7 mm × 0.031 = 3 mm]
Force = 28,910 × 1.031 = 29,806 lbs [128.6 kN × 1.031 = 132.6 kN]

Total Gross Theoretical Load Required = 29,806 + 1,495 = 31,301 lbs [132.6 + 6.97 = 139.6 kN]
This load is less than 80% of the ultimate strand strength (33,000 lbs [146.8 kN]), therefore it is not necessary to jack at both anchorages.

Total Net Theoretical Load Required = 29,806 lbs [132.6 kN]

Elongation Computation Summary:

	Gross Theoretical Elongation		Net Theoretical Elongation	
Basic Elongation	3.25 in.	82.7 mm	3.25 in.	82.7 mm
Dead End Seating Loss	0.0625	1.5	0.0625	1.5
Live End Seating Loss	0.1875	5	0.0	0
Friction Losses	0.101	3	0.101	3
Total Elongation	3.601 in.	92.2 mm	3.4135 in.	87.2 mm
Rounded	3⅝ in.	92 mm	3⅜ in.	87 mm
Tolerance Limits	-5% = 3⅜ in. [87.4 mm]		-5% = 3¼ in. [82.7 mm]	
	+5% = 3¾ in. [96.6 mm]		+5% = 3⅝ in. [91.4 mm]	

Use Gross Theoretical Elongation for monitoring travel of ram, and compare to 31,303 lb [139.6 kN] force. Use Net Theoretical Elongation for comparison against movement of mark on strand from initial tension reference.

C. Post-tensioned Panel Using Looped Single Strand Tendon

The following example details the method for calculating the elongation of a looped, greased and plastic coated (unbonded) strand. Adjustments for anchor wedge seating, and friction losses are shown. Note that with the exception of including the curvature friction loss factor, the procedure for calculating the tensioning parameters is the same as for Sample B.

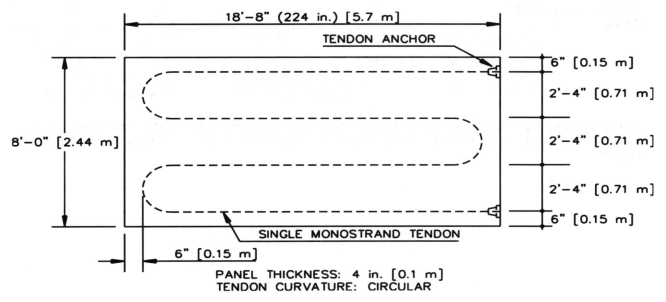

First, the necessary material data and tensioning setup information must be established.

1. Size and type of strand: 1/2 in. [12.7 mm] diameter, 270K [1,860 MPa].
2. Physical characteristics of strand:
   From the mill certificate supplied by the manufacturer,
   $A = 0.1532$ in.^{2} [98.84 mm^2]
   $E = 28,650,000$ psi [197,542 MPa]
   The average values being used by the plant are,
   $A = 0.1530$ in.^{2} [98.71 mm^2]
   $E = 28,500,000$ psi [196,500 MPa]
   Comparing the actual values to the averages,

   $$\frac{AE}{A_a E_a} = \frac{0.1532 \times 28,650,000}{0.1530 \times 28,500,000} \times 100 = 100.7\%$$

   $$\left[\frac{98.84 \text{ mm}^2 \times 197,542 \text{ MPa}}{98.71 \text{ mm}^2 \times 196,500 \text{ MPa}} \times 100 = 100.7\% \right]$$

   Since AE is within 2.5% of average, the average values can be used in the calculations. *Note that if AE was not within 2.5% of average, this does not indicate that the strand is not in compliance with specification, but only that the actual values from the mill certificate must be used in calculations.*

3. From information supplied by the tendon manufacturer,
   κ (wobble friction coefficient) = 0.0007 per foot of tendon [0.0023 per meter]
   μ (curvature friction coefficient) = 0.05

Corrections to Tensioning:

a. Dead End Wedge Seating Loss
   Based on ongoing monitoring, seating after initial tension is applied is expected to be 1/8 in. [3 mm].

b. Live End Wedge Seating Loss
   Expect 1/8 in. [3 mm] based on past history. Over pull of 1/8 in. [3 mm] is required.

c. Elastic Shortening of Panel
   Neglect effects of elastic shortening for this short panel.

4. Use initial tension of 3,000 lbs [13.344 kN]

5. Strand is to be stressed to 70% of ultimate,

   41,300 lbs x 0.70 = 28,910 lbs

   [183.7 kN x 0.70 = 128.6 kN]

Friction Losses:

Friction in the tendon system will result in a reduced strand stress at the dead (non-jacking) end of the tendon. Thus, some over pull is required to ensure that the average strand stress equals the design value. When friction losses are high, it is recommended that sequential jacking at both ends of the tendon be used to reduce the possibility of overstressing the strand at the live end.

$$\frac{P_D}{P_S} = e^{-(\kappa L + \mu\alpha)}$$

where P_D equals the force in the strand at the dead end, P_S equals the force in the strand at the live end, L is the tendon length, in feet, between anchorages, and α is the total angular change of the tendon profile, in radians, between the anchorages. For this example,

Tendon Length = (224 in. − 6 in. − 14 in.) (2) + (224 in. − 12 in. − 28 in.) (2) + (π x 14 in. x 3) = 908 in.
[(5.7 m − 0.15 m − 0.355 m) (2) + (5.7 m − 0.305 m − 0.71 m) (2) + (π x 0.355 m x 3) = 23.1 m]

Curvature = π(3) = 9.42 radians

$$\frac{P_D}{P_S} = e^{-(0.0007 \times 75.67 \text{ ft} + 0.05 \times 9.42)} = 0.59$$

$$\left[e^{-(0.0023 \times 23.1 \text{ m} + 0.05 \times 9.42)} = 0.59 \right]$$

Average strand force = (1 + 0.59)/2 = 0.795, loss = 20.5%. This will require jacking from both ends of the tendon to compensate for friction losses.

Calculate the friction loss at mid-point of the tendon,

$$\frac{P_M}{P_S} = e^{-\left(\frac{0.0007 \times 75.67 \text{ ft} + 0.05 \times 9.42}{2}\right)} = 0.77$$

$$\left[e^{-\left(\frac{0.0023 \times 23.1 \text{ m} + 0.05 \times 9.42}{2}\right)} = 0.77 \right]$$

Average strand force over one half of the tendon = (1 + 0.77)/2 = 0.885, loss = 11.5%

$$\text{Basic Elongation} = \frac{(\text{Force required beyond initial tension})(\text{Length of strand between anchorages})}{(\text{Area of strand})(\text{Modulus of elasticity})}$$

$$\text{Basic Elongation} = \frac{(28,910 - 3,000) \text{ lbs} \times 908 \text{ in.}}{0.153 \text{ in.}^2 \times 28,500,000 \text{ psi}} = 5.40 \text{ in.}$$

$$\left[\frac{(128.6 - 13.3) \text{ kN} \times 23.1 \text{ m}}{(98.71 \times 10^{-6} \text{ m}^2 \times 196,500 \text{ MPa} \times 1000)} = 0.137 \text{ m} \right]$$

Theoretical Elongation = Basic Elongation combined with appropriate corrections.

a. Dead End Seating : Add 1/8 in. [3 mm] to elongation. No adjustment to force is required.

b. Live End Seating: Over pull by 1/8 in. [3 mm]. Adjust force accordingly.

$$\frac{5.40 \text{ in.}}{25{,}910 \text{ lbs}} = \frac{0.125 \text{ in.}}{\text{Added Force}} \quad \text{therefore,} \quad \text{Added Force} = \frac{0.125 \times 25{,}910}{5.40} = 600 \text{ lbs}$$

$$\left[\frac{(3 \text{ mm}) \times (115.3 \text{ kN})}{137 \text{ mm}} = 2.53 \text{ kN} \right]$$

c. Neglect Elastic Shortening

d. Friction Losses
We would exceed the breaking strength of the strand if total friction losses were included in the tensioning force. Tension at one anchorage to the Gross Theoretical Values and note the elongation achieved. Jack at the other anchorage to the same force value and again note the elongation. Add the two elongation values to obtain the total. This total must agree with the calculated value within 5%.

Elongation Adjustment = 5.40 x 0.115 = 0.621 in. [137 mm x 0.115 = 15.8 mm]
Force = 28,910 x 1.115 = 32,234 lbs [128.6 kN x 1.0115 = 143.4 kN]

Total Gross Theoretical Load Required = 32,234 + 600 = 32,834 lbs [143.4 + 2.53 = 145.9 kN]
This load is less than 80% of the ultimate strand strength (33,000 lbs [146.8 kN]), therefore our procedure is acceptable. Note that if the result was greater than 33,000 lbs, a multi-stage tensioning procedure would be required with incremental tensioning steps at alternate anchorages until the total Gross Theoretical Elongation was achieved.

Elongation Computation Summary:

	Gross Theoretical Elongation		Second Stage Elongation	
Basic Elongation	5.40 in.	137 mm	5.40 in.	137 mm
Dead End Seating Loss	0.125	3	0.0	0
Live End Seating Loss	0.125	3	0.125	3
Friction Losses	0.621	15.8	0.621	15.8 mm
Total Elongation	6.271 in.	158.8 mm	6.146 in.	155.8
Rounded	6¼ in.	159 mm	6⅛ in.	156 mm
Tolerance Limits	-5% = 6 in. [151 mm] +5% = 6⅝ in. [167 mm]		-5% = 5⅞ in. [148 mm] +5% = 6½ in. [164 mm]	

For the initial tensioning stage use Gross Theoretical Elongation for monitoring travel of ram, and compare to 32,834 lb [143.4 kN] force. At second stage tensioning, the force is the same, but the elongation measurement is reduced by the "dead end seating loss", since the strand has already seated at the opposing chuck.

APPENDIX I

Erection Tolerances

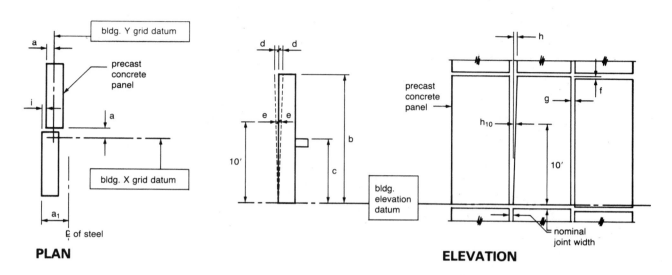

Precast element to precast or cast-in-place concrete, masonry, or structural steel

a = Plan location from building grid datum*..........................±1/2 in. (±13 mm)
a_1 = Plan location from centerline of steel **..........................±1/2 in. (±13 mm)
b = Top elevation from nominal top elevation

    Exposed individual panel..........................±1/4 in. (±6 mm)
    Nonexposed individual panel..........................±1/2 in. (±13 mm)
    Exposed relative to adjacent panel..........................1/4 in. (6 mm)
    Nonexposed relative to adjacent panel..........................1/2 in. (13 mm)

c = Support elevation from nominal elevation

    Maximum low..........................1/2 in. (13 mm)
    Maximum high..........................1/4 in. (6 mm)

d = Maximum plumb variation over height of structure or
100 ft (30 m) whichever is less*..........................1 in. (25 mm)

e = Plumb in any 10 ft (3 m) of element height..........................1/4 in. (6 mm)
f = Maximum jog in alignment of matching edges..........................1/4 in. (6 mm)
g = Joint width (governs over joint taper)..........................±1/4 in. (±6 mm)
h = Joint taper maximum..........................3/8 in. (9 mm)
h_{10} = Joint taper over 10 ft (3 m)..........................1/4 in. (6 mm)
i = Maximum jog in alignment of matching faces..........................1/4 in. (6 mm)
j = Differential bowing or camber as erected between
adjacent members of the same design..........................1/4 in. (6 mm)

---

* For precast buildings in excess of 10 ft (30 m) tall, tolerances "a" and "d" can increase at the rate of 1/8 in. (3 mm) per story over 100 ft (30 m) to a maximum of 2 in. (50 mm).

** For precast elements erected on a steel frame, this tolerance takes precedence over tolerance on dimension "a".

APPENDIX J

Architectural Trim Units Category

Architectural Precast Concrete Trim Units (AT) is a category of products for certification within the Architectural Products Group. This category is defined as,

> Non-prestressed products with an architectural surface finish and of relatively small size that can be installed with equipment of limited capacity, such as bollards, benches, planters, and pavers, and cast stone building units such as sills, lintels, coping, cornices, quoins, and medallions.

This definition includes specialty products specifically designed for a project, and not commodity or ornamental units produced for wholesale or retail purposes. This category does not include dry-cast products such as machine made pavers, or dry-tamped cast stone.

> For this class of products, the criteria established in this manual (MNL-117) for quality control govern except as specifically noted below. The section numbers listed correspond to sections in the manual.

Standard

1.3 Personnel

1.3.2 Engineering

Plants shall have available the services of a registered professional engineer experienced in the design of precast concrete. The engineer shall prescribe design policies and be competent to review designs prepared by others. The engineer shall be responsible for the design of all products for production and handling.

1.3.5 Quality Control

Quality control procedures shall be established by the plant's general management, with duties assigned to qualified personnel such that authority for carrying out the responsibilities of quality control is maintained by general management or engineering, and is not a function of the production staff.

Responsibilities shall include assuring that the following activities are performed at a frequency shown to be adequate to meet quality objectives or as prescribed in this manual.
   a. Inspecting and verifying the accuracy of dimen sions and condition of molds.
   b. Verifying batching, mixing, material handling, placing, consolidating, curing, product handling and storage procedures.
   c. Verifying the proper fabrication and place -

Commentary

C1.3 Personnel

C1.3.2 Engineering

Engineering personnel should review the design of precast concrete elements prepared by the engineer of record. The engineer should have the ability to solve problems and devise methods, as required, for the design, production, handling, and/or erection of precast concrete products.

C1.3.5 Quality Control

Due to the limited staff at most plants producing Architectural Trim Units as their primary product, it is impractical to require a separate quality control department in the organizational structure. The intent of this section is to assure that authority for maintaining quality requirements is not vested with personnel who are primarily concerned with productivity.

The qualifications of personnel conducting inspections and tests are critical to providing adequate assurance that the precast concrete products will satisfy the desired level of quality.

All personnel should observe and report any changes in plant equipment, working conditions, weather and other items which have the potential for affecting the quality of products.

Standard

ment of reinforcement, and quantity and location of cast-in items.

d. Preparing or evaluating mix designs.
e. Taking representative test samples and performing all required testing.
f. Inspecting all finished products for conformance with shop drawings, approved samples and project requirements.
g. Preparing and maintaining complete quality control records.

3.2 Reinforcement and Hardware

3.2.2 Prestressing Materials

This section does not apply to Category AT.

3.3 Insulation

This section does not apply to Category AT.

4.1 Mix Proportioning

4.1.1 Qualification of New Concrete Mixes

Concrete mixes for precast concrete shall be established initially by laboratory methods. The proportioning of mixes shall be done either by a qualified commercial laboratory or qualified precast concrete plant personnel. Mixes shall be evaluated by trial batches prepared in accordance with ASTM C192 and plant tests under conditions simulating as closely as possible actual production and finishing.

Each concrete mix used shall be developed using the brand and type of cement, source and gradation of aggregates, and the brand of admixture proposed for use in the production mixes. If any of these variables are changed, the proportions of the mixture shall be re-evaluated.

Where a history of use of a concrete mix with similar proportions and materials to those of the proposed production mix is available, laboratory testing and evaluation is not required.

Concrete mixes shall be proportioned and/or evaluated for each individual project with respect to

Commentary

C4.1 Mix Proportioning

C4.1.1 Qualification of New Concrete Mixes

Due to the large variety of concrete mixes used for relatively small volume production runs, it is impractical to require separate trial batches for all mix designs. Initial mix proportions can be based on previously used concrete mix designs for which sufficient data is available to predict strength, durability, and workability of the new design. Tests for strength and air content should be conducted during the preparation of color and texture samples to confirm the applicability of the reference concrete mix.

Standard

strength, absorption, volume change, and resistance to freezing and thawing where such environments exist, as well as desired surface finish (color and texture). Mixes shall have adequate workability for proper placement and consolidation.

5.2 Prestressing

This section does not apply to Category AT.

5.3 Pretensioning

This section does not apply to Category AT.

5.4 Post-Tensioning of Plant-Produced Products

This section does not apply to Category AT.

6.2 Testing

6.2.3 Production Testing

The requirements for frequency of testing during production are modified for Category AT products as follows.

1. Aggregates
   A sieve analysis (ASTM C136) and unit weight test (ASTM C29) shall be conducted in the plant with test samples taken at any point between and including stockpile and batching hopper. Such tests shall be carried out for each aggregate type and size at least once every 4 weeks, or for every 40 cu. yds. (31 m^3) of an aggregate used in a 4 week period when usage in that period exceeds 40 cu. yds. (31 m^3).

2. Concrete Strength
   For each concrete mix in use, a minimum of four compression test specimens, made in accordance with ASTM C31, shall be made weekly or for every 15 cu. yds. (11.5 m^3) produced, whichever is more frequent. Two of these specimens shall be used to verify 28-day design strength, and at least one of the specimens shall be used to verify concrete strength at the maturity being used for removal of products from their forms.

   Two compression specimens shall be made daily for each individual concrete mix used. These

Commentary

C6.2 Testing

C6.2.3 Production Testing

Due to the relatively low volume of materials used in production of category AT products, the minimum testing frequency requirements are modified to levels more reasonable for the typical production cycles.

2. Concrete Strength

   When small volumes of concrete are used, and data is available to support the anticipated strength of a concrete mix design, a reduced frequency of testing is appropriate. At least once each week, or for every 15 cu. yds., (11.5 m^3) a set of tests should be conducted to verify concrete mix design performance. This is a minimum requirement. Any indication of potential for reduced concrete performance is cause for increased frequency of testing.

   The use of 4 x 8 in. (100 x 200 mm) cylinder specimens is recommended.

   The two test specimens made each day may be tested at 28-days, or at any age deemed appropriate for the appli-

Standard

specimens shall be made in accordance with ASTM C31, except that a "Chace Indicator" test, in accordance with AASHTO T-199, may be used to determine air content.

4. Air Content
For periodic daily checks, a "Chace Indicator" test, in accordance with AASHTO T-199, may be used.

6.2.4 Special Testing

1. Heat of Hydration
Testing is not applicable to Architectural Trim Units category.

7.1 Requirements for Finished Product

7.1.2 Product Tolerances

1. Sills, Lintels, Coping, Cornices, Quoins, Medallions
   a. Overall height and width of units, measured at the face(s) exposed to view . . . ±1/8 in. (3mm)
   b. Total thickness of flange thickness
   . . . . ±1/8 in. (3mm)
   Where one face will be installed in dead wall space or mortar joint . . . . . +1/4 in. (6mm), maximum

2. Bollards, Benches, Planters
   . . . . ±1/4 in. (6 mm), all dimensions

3. Pavers
   a.   Width and thickness
   . . . . +1/16 in. (1.5 mm), -1/8 in. (3 mm)
   b.   Length 2 ft. (0.6 m) or less
   . . . . +1/16 in. (1.5 mm), -1/8 in. (3 mm)
   2 to 5 ft (0.6 to 1.5 m) . . . . ±1/8 in. (3 mm)
   5 to 10 ft. (1.5 to 3 m)
   . . . . +1/8 in. (3 mm), -3/16 in. (4.5 mm)

4. Size and location of rustications and architectural features . . . . ±1/16 in. (1.5 mm)

5. Location of inserts and appurtenances:
   a. on formed surfaces . . . . ±1/8 in. (3 mm)
   b. on un-formed surfacaes . . . . . ±3/8 in. (9 mm)

6. Bowing . . . . L/360, 1/8 in. max (6 mm)

7. Local smoothness . . . . ±1/8 in. in 5 ft. (3 mm in 1.5 m)

Commentary

cation (such as prior to shipping). These specimens may also be used to verify quality of production runs if the weekly test set indicates a potential problem.

4. Air Content
The "Chace Indicator" will give an approximate value for the air content. Any indication that the air content is not within specified limits would be cause to test in accordance with ASTM C173 or C231.

C6.2.4 Special Testing

1. Heat of Hydration
This class of products does not involve "massive castings".

C7.1 Requirements for Finished Product

C7.1.2 Product Tolerances

These tolerances are the minimum acceptable criteria in the absence of specified requirements. Project specifications, or product application, may require more stringent tolerances.
Refer to Section 7.1.2 for tolerances not listed here.

8. Warping:
   a. Pavers
   ..... The numerically greater of 1/16 in. (1.5 mm) or 1/32 in. per foot, (0.75 mm per 300 mm) of distance from nearest adjacent corner.
   b. Other products
   ...... 1/16 in. per foot (1.5 mm per 300 mm) of distance from nearest adjacent corner, 1/4 in. maximum (3 mm)

INDEX

Term	Section Number
Absorption Test	6.2.2 (10)
Accelerated Curing	2.2.6
General	4.21.1
Live Steam	4.21.2
Radiant Heat & Moisture	4.21.3
Temperature	4.21.1
Acceptability	1.5.4
Appearance	2.9
Acid Etching	2.7.4
Admixtures	3.1.7
Accelerating	3.1.7
Air Entraining	3.1.7
Batching	4.16.5
Certification	6.2.2 (5)
Coloring	3.1.7
Dispensers	4.9.4
Effects of	4.8
High Range Water Reducer	3.1.7, 4.6
Mineral	3.1.7
Retarding	3.1.7
Storage	4.9.4
Aggregate	
Batching	4.16.2
Exposure	2.7.1
Gradation	3.1.3, 6.2.2 (2), 6.2.3 (1)
Moisture Content	4.16.2
Segregation	4.18.3
Segregation Lines	2.9
Storage	4.9.2
Temperature	C3.1.2
Testing	6.2.2 (2), 6.2.3 (1)
Transparency or Shadowing	C2.7.2
Aggregates, Backup	3.1.4
Aggregates, Facing	3.1.3
Air Content	
Testing	6.2.2 (10), 6.2.3 (4)
Air Entrainment	4.2, C4.17.3
Air Temperature	6.2.3 (7)
Alkali-Carbonate Reaction	
Testing	6.2.2 (2)
Alkali-Silica Reaction	
Testing	6.2.2 (2)
Anchorage	
Protection	5.3.14

Term	Section Number
Architectural Trim Units	Appendix J
Backup Concrete	
Aggregates	3.1.4
Bleeding	2.9
Consolidation	4.19.3
Placing	4.18.10
Backup Mix	C4.1, 4.3
Bar Supports	3.2.1, 5.1.4
Batch Plant	4.9.1
Batchers, General	4.13.1
Batching	
Admixtures	4.16.5
Aggregates	4.16.2
Cement	4.16.3
General	4.16.1
Lightweight Aggregates	4.17.7
Pigment	4.16.3
Water	4.16.4
Batching Equipment	
Tolerances	4.10
Blocking Stains	2.9
Bond Breakers	2.7.9
Brick	2.7.9
Bugholes	2.9, C2.7.2, C2.7.3
Cage Assemblies	5.1.3
Calibration Records	6.3.5
Gauging Systems	5.2.4
Jacking Equipment	5.2.7
Cement	3.1.2
Batching	4.16.3
Lumps	4.9.3
Mill Certificates	6.2.2 (1)
Storage	4.9.3
Temperature	C4.18.7
Chlorides	3.1.7
Chuck Seating	5.3.10
Chucks	3.2.2
Splice	5.3.5
Strand	5.3.5
Use and Maintenance	Appendix D
Cleaning	2.6.4
Coarse Aggregates	3.1.3

Term	Section Number
Size	3.1.3
Weighing	4.11
Coatings	2.7.12
Cold Weather, Mixing	4.17.8
Color	
Uniformity	2.9, 3.1.2, 3.1.7, 4.6
Compressive Strength	
Testing	6.2.2 (10), 6.2.3 (2)
Concrete	
Handling	2.2.5
Temperature	2.2.6, 6.2.3 (6)
Concrete Mixtures	6.2.2 (10)
Concrete Records	6.3.4
Concrete Temperature	4.18.5, 4.18.7, 4.18.8
Concrete Transportation Equipment	
Agitating	4.14.2, 4.18.2
General	4.14.1
Consolidation	
Backup Mixes	4.19.3
External Form Vibrators	4.19.5
Face Mixes	4.19.3
Flowing Concrete	C4.19.4
General	4.19.1
Internal Vibrators	4.19.4
Lightweight Concrete	4.19.2
Surface Vibrators	4.19.6
Vibrating Tables	4.19.7
Corrosion Protection	
Hardware	3.2.3
Lifting Devices	3.2.4
Cover of Reinforcing Steel	5.1.4
Cracks	C2.8, 2.9
Crazing	C2.7.2, C2.9
Curing	2.2.6
Facilities	2.2.6
General	4.20.1
Membrane Curing Compound	4.22.3
Moist	4.22.1
Moisture Retention Enclosures	4.22.2
Temperature	4.20.2
Design Responsibilities	Appendix B
Detensioning	5.3.13
Dissimilar Metals	2.5, 3.2.3
Dunnage	2.6.3
Durability	4.1.4

Term	Section Number
Efflorescence	C3.1.7
Electrodes	3.4
Elongation	
Calculations	5.3.10
Corrections	5.3.10
Measurement	5.3.9
Strand	3.2.2
Epoxy Coated Reinforcement	5.1.2
Bar Supports	5.1.4
Repair	5.1.3
Face Mix	C4.1, 4.3
Consolidation	4.19.3
Gap-graded	6.2.2 (2)
Placing	4.18.9
Thickness	4.18.9
Facing Aggregates	3.1.3
False Set	C4.17.9
Fine Aggregates	3.1.3
Weighing	4.11
Finish	
Acceptability	1.5.4
Samples	Appendix C
Finishes	
Acid Etched	2.7.4
Clay Product Veneer Facing	2.7.9
Form Liner	2.7.8
Honed or Polished	2.7.7
Retarded	2.7.5
Sand Embedded Materials	2.7.10
Sand or Abrasive Blast	2.73
Smooth	2.7.2
Stone Veneer Facing	2.7.9
Tooled or Bushhammered	2.7.6
Unformed Surface	2.7.11
Finishing Areas	2.2.6
Flash Set	4.17.8
Floating	2.7.11
Fly Ash	C3.1.7
Force Corrections	5.3.11
Force Measurement	5.2.2, 5.2.3
Tolerance	5.3.8
Form Release Agents	6.2.2 (9)
Form Shortening	5.3.10
Forms	
Records	Appendix E
Freeze-Thaw Resistance	4.1.4, 4.2

Term	Section Number
Freeze-Thaw Tests	6.2.4 (2)
Galvanized Hardware	3.2.3
Galvanized Reinforcement	3.2.1
Bar Supports	5.1.4
Repair	5.1.3
Galvanizing	2.3.1
Gauging Systems	5.2.4
Grout	3.2.2
Handling	
Product	2.6.1
Hardware	3.2.3
Anchorage	2.5
Certification	6.2.2 (7)
Fabrication	2.2.4
Galvanized	3.2.3
Installation	2.5
Materials	3.2.3
Storage	2.2.4
Heat of Hydration	6.2.4 (1)
Hot Weather	
Mixing	4.17.9
Inserts	
Certification	6.2.2 (7)
Nonferrous	3.2.3
Wood	3.2.3
Inspection	
Post-pour	6.1.2
Pre-pour	6.1.2
Insulation	3.3
Jacking Force	
Allowable	5.3.12
Lifting	
Devices	3.2.4
Loops	2.5
Lightweight Aggregates	3.1.5, 4.3
Batching	4.17.7
Consolidation	4.19.2
Mix Proportioning	4.5
Loading	2.2.7, 2.6.5
Lumber	2.5
Metakaolin	3.1.7
Batching	4.16.5
Curing	4.20.1
Storage	4.9.4
Mix Proportioning	4.1, 4.1.1, 4.4, 4.1.4, 4.5, 4.6
Mixers	
Maintenance	4.13.4
Requirements	4.13.2, 4.13.3
Mixing	
Cold Weather	4.17.8
General	4.16.1, 4.17.1
Hot Weather	4.17.9
Methods	4.17.2
Shrink Mixing	4.17.5
Stationary Mixers	4.17.4
Time	4.17.3
Truck Mixing	4.17.6
Mixing Water	C3.1.2, 3.1.6
Mockup Process	2.9
Mockups	1.5.4
Moisture Meters	4.16.2
Mold	2.2.2, 2.4.1, 2.4.2
Beds	C2.2.2
Construction	2.4.1
Envelope	C2.4.1
Fabrication	2.2.2
Master	C.2.4.1
Materials	2.4.1
Plastic	2.4.1
Preparation	4.18.4
Steel	2.4.1
Nonferrous Metals	3.2.3
Paint	2.7.12
Pigments	3.1.7
Batching	4.16.3
Certification	6.2.2 (6)
Storage	4.9.4
Placing	
Aggregate Segregation	4.18.3
Backup Concrete	4.18.10
Cold Weather Conditions	4.18.8
Concrete Temperature	4.18.5, 4.18.7, 4.18.8
Equipment	4.15
Facing Concrete	4.18.9
General	4.18.1
Hot and Windy Conditions	4.18.7
Severe Weather Conditions	4.18.5
Wet and Rainy Conditions	4.18.6
Plant Certification Program	Appendix F
Plant Facilities	2.2.1, 2.2.5
Plant Manual	1.2.1, 1.2.2, Appendix A
Plastic Shrinkage Cracking	4.8.7, 4.20.1
Post Tensioning	
Bars	3.2.2

Term	Section Number
Ducts	5.4.2
Elongation Measurement	5.4.1
Final force	5.4.4
Friction in Ducts	5.4.3
Grouting	5.4.6
Initial force	5.4.4
Records	5.4.1
Sealing	5.4.7
Sheathing	3.2.2
Strand	3.2.2
Stress Measurement	5.4.1
Tensioning	5.4.4
Unbonded Tendons	5.4.1
Wire	3.2.2
Post Tensioned Anchorages	3.2.2, 5.4.5
Prestressing Steel	
Corrosion	5.3.1
Mill Certificates	6.2.2 (4)
Rust	3.2.2
Storage	5.3.1
Strand Surfaces	5.3.3
Quality Control	1.3.1
Personnel	1.2.3, 1.3.5
Records	1.2.2
Responsibilities	1.3.5
Recordkeeping	6.3.1
Calibration Records	6.3.5
Concrete Records	6.3.14
Forms	Appendix E
Supplier's Tests	6.3.2
Tensioning Records	6.3.3
References	Appendix G
Reinforcement	
Epoxy Coated	3.2.1, 4.19.4
Fabrication	5.1.3
Galvanized	3.2.1
Identification	C2.3.2
Installation	5.1.4
Mill Certificates	6.2.2 (4)
Rust	3.2.1
Shadow Lines	2.9
Splices	5.1.3
Storage	5.1.2
Welded Wire	3.2.1
Welding	2.3.2, 3.2.1
Release Agents	2.2.3, C2.2.5, 2.4.1
Release Strength	4.1.2
Curing	4.20.3
Repairs	2.8

Term	Section Number
Retarded Surface	2.7.5
Retarders	6.2.2 (9)
Retempering	4.18.2
Returns	C2.7.1, 2.9
Safety	2.1.2
Samples	1.5.1, 1.5.2, 1.5.3, 2.9
Finish	Appendix C
Mockups	Appendix C
Pre-Bid	Appendix C
Sand Embedment	2.7.10
Scale Requirements	
Tolerances	4.11
Sealer	2.2.6, 2.4.1, 2.10, 6.2.2 (9)
Sheathing	3.2.2
Shop Drawings	1.3.3, 1.4.2, 3.2.4
Silica Fume	3.1.7, C4.7.1
Batching	4.16.5
Curing	4.20.1
Storage	4.9.4
Slump	3.1.7, 4.6, 4.17.3
Testing	6.2.2 (10), 6.2.3 (4)
Staining	
Water	3.1.6
Staining (iron sulfides)	6.2.2 (2)
Stainless Steel	2.2.4, 3.2.3
Hardware	2.3.1
Installation	2.5
Studs	C2.3.3
Stains	2.6.4
Blocking	C2.9
Rust	2.9
Storage	
Finished Products	2.2.8
Hardware	2.2.4
Product	2.6.3
Release Agents and Retarders	2.2.3
Storage Racks	2.2.7, 2.2.8, 2.6.3
Strand	3.2.2
Elongation	3.2.2
Position	5.3.7
Protection of Ends	5.3.14
Splices	5.3.6
Strength (of Concrete)	4.1.2
Design, Stripping, Transfer	4.1.3
Stressing of Strands	
Final	5.3.12

Term	Section Number
Stressing Procedure	5.3.2
Stringing of Strands	5.3.4
Stripping	2.6.2, 4.1.2, 4.20.3
Structural Steel	3.2.3
Stud Welding	2.3.3
Certification	6.2.2 (8)
Inspection	6.2.3 (9)
Vendor Testing	6.2.2 (8)
Studs	2.3.3
Supplier's Test Reports	6.3.2
Surface Finishes (see Finishes)	2.7
Tack Welds	2.3.2
Temperature	
Concrete	2.2.6
Curing	4.20.2
Measuring	2.2.6
Tensioning	5.2.2
Calculations	Appendix H
Force Application	5.3.2
Force Measurement	5.2.4
Initial	5.3.8
Jacking Force Control	5.2.5
Records	6.3.3
Terra Cotta	2.7.9
Testing Equipment	6.4.2
Operating Instructions	6.4.3
Texture	
Surface Finish	2.7.1
Variations	2.9
Tile	2.7.9
Tolerances	
Erection	Appendix I
Product	7.1.1, 7.1.2
Transfer Strength	4.1.2
Transporting Concrete	4.18.2
Agitating Equipment	4.14.1, 4.14.2
Unformed Surface Finishes	2.7.11
Uniformity	C2.7.1, 3.1.1
Uniformity of Concrete	4.17.3
Unit Weight, Testing	6.2.2 (10), 6.2.3 (5)
Vibration (See Consolidation)	
Vibrators	2.2.5
Water, Testing	6.2.2 (3)
Water Absorption, Testing	6.2.2 (10)

Term	Section Number
Water Measuring Equipment	4.12
Water-Cement Ratio	4.7, 4.7.1, 4.7.2, 4.7.3
Welded Wire Reinforcement	3.2.1, 5.1.2
Bending	5.1.3
Inspection	6.2.2 (4)
Mill Certificates	6.2.2 (4)
Welder	
Qualification	2.3.1
Welding	2.2.4, 2.3.2
Bar Identification Marks	C2.3.2
Carbon Equivalent	2.3.2
Electrodes	2.2.4, 2.3.1, 2.3.2, 3.4
Galvanized Steel	2.3.1, 2.3.2
Headed Studs	3.2.3
Inspection	6.2.3 (8)
Mill Scale	2.3.2
Preheat	2.3.1, 2.3.2
Reinforcement	2.3.2
Reinforcing Bars	3.2.1
Stainless Steel	2.3.1
Structural Steel	2.3.1
Tack Welding	2.3.2, 2.3.3
Welding Slag	2.3.2
Wire Failure in Strand	5.2.6
Zinc Coated Reinforcement (see Galvanized Reinforcement)	
Zinc Rich Paint	2.3.1